MINERALS
OF NORTHERN ENGLAND

R. F. Symes and B. Young

Published in 2008 by
NMS Enterprises Limited – Publishing
a division of NMS Enterprises Limited
National Museums Scotland
Chambers Street
Edinburgh EH1 1JF

in association with the
Natural History Museum
Cromwell Road
London SW7 5BD

ISBN (10): 1 905267 01 0
ISBN (13): 978 1 905267 01 9

The rights of R. F. Symes and B. Young to be
identified as the authors of this book have been
asserted by all in accordance with the Copyright,
Designs and Patents Act 1988.

British Library Cataloguing in Publication Data
A catalogue record of this book
is available from the British Library.

Cover design by Mark Blackadder.
Cover images: (front) Fluorite [BM 1964, R 1524],
    see page 152; (back) Calcite [BM 1926, 1252],
    see page 137; both images ©NHMPL.
Internal layout by NMS Enterprises Limited –
    Publishing.
Printed and bound in the United Kingdom
    by Cambridge Printing.

For a full listing of titles and related merchandise,
please contact:

www.nms.ac.uk/books

www.nhmshop.co.uk

Image Credits

Contents

Foreword

NORTHERN England, an area of great natural beauty, is additionally famous for numerous, and varied, orefields. The latter rank as one of the most outstanding features of British geology and are world-renowned. From a mining perspective, coupled with a remarkable wealth of classic, exquisite mineral specimens, the area both rivals, and compliments, Cornwall in the annals of British mineralogy.

The authors have accessed extensive mining and mineralogical literature to integrate very successfully a comprehensive geological and mineralogical understanding of the orefields, the minerals, and their origin. In their mammoth task of compilation they include mines, and minerals throughout the Lake District, the north Pennines, Cumbria, various regions of north-east England, and North Yorkshire. Such a diverse expanse naturally leads to a mineral richness which generated half of the known mineral species in Britain. A most remarkable feature of the Alston Block is a concentric zoning of constituent minerals which makes it one of the world's finest examples of a zoned orefield. Indeed, the north Pennines are designated a Global Geopark. Economic activity additionally spawned trade in minerals, mineral collections and dealers, local museums and, more recently, heritage site development. Noteworthy mineral collectors include John Woodward, Philip Rashleigh, Henry Ludlam and Sir Arthur Russell.

Throughout the work frequent mention is made of the current situation on the ground; trial workings, mine dumps, mine buildings and smeltings, plus mine location maps and a wealth of historic photographs all bring the mining history alive to the reader. Some mining companies took great interest in their employees' welfare, for welfare schemes, social amenities and education systems were set up: thus the narrative humanises the miners' arduous life in their day-to-day harsh existence. Aspects of social and welfare inter-action generated many agricultural and domestic features currently visible in the north Pennines landscape. Miners' interests, and creativity, is also reflected in the wonderful, alluring spar towers, spar boxes and grottos they constructed.

Minerals are the main focus of the book. As a measure of their display-cabinet quality – especially fluorite, barite, hematite and witherite – they are to be found in major world-wide collections. Mineral species and locality data smoothly integrate into the numerous descriptions of the mines. A great strength of this work is the high number of superb, large to full-plate, images of the minerals. Photographs of stunning barite crystals, calcite, beautiful calcite butterfly twins, witherite, alstonite, hematite, cerussite, and of course the world-famous fluorite, plus other species, bring an almost 'tactile, three-dimensional hand specimen experience' to the reader as they behold such objects of great beauty. A selected bibliography concludes the work.

Minerals of Northern England will rate as one of the classic works on British mineralogy for many years to come.

Dr Alec Livingstone
Former Keeper of Minerals
NATIONAL MUSEUMS SCOTLAND, EDINBURGH

Acknowledgements

THE authors are extremely grateful to many people, friends and colleagues in numerous organisations who have assisted in various ways during the compilation of this book. In a work of this sort it is impossible to name all who have helped, although special thanks must be accorded to the following. Numerous individuals and organisations have allowed the use of photographs: these are acknowledged separately.

Staff of the Department of Mineralogy at the Natural History Museum (NHM), London, in particular Dr Chris Stanley and Peter Tandy, gave invaluable assistance in the selection of specimens for photography and in providing background information on them. Field photography, and photography of specimens in London was undertaken by Frank Greenaway and Harry Taylor of the Photographic Unit, NHM. At the National Museums Scotland, Edinburgh, Alec Livingstone and Brian Jackson assisted in specimen selection. Specimen photography in Edinburgh was carried out by Suzie Stevenson of the National Museums Scotland. For access to collections in their care,

loan of specimens and facilities to undertake photography, we thank Hazel Davidson of Keswick Museum, Stephen Hewitt of Tullie House Museum, Carlisle, Ian Forbes of Killhope Lead Mining Museum, Weardale, and the Friends of Killhope, Professor Bob Holdsworth of the Department of Earth Sciences, University of Durham, David and Elizabeth Hacker, Maurice Wall and Alan Pringle. We also wish to thank Michael Cooper, of Nottingham City Museums for information and advice on collectors and dealers, and Ian Forbes of Killhope Lead Mining Museum for his advice on the history of mining in the northern Pennines.

Particular thanks go to Alec Livingstone, formerly of National Museums Scotland, for his role in bringing the text for this book to the notice of the National Museums Scotland and for his generous encouragement throughout the project.

This list would not be complete without acknowledging our gratitude to Lesley Taylor and Marián Sumega, for their skill, enthusiasm and patience with the authors throughout the project.

R. F. Symes and B. Young

About the Authors

Dr R. F. (Bob) Symes, OBE

Bob Symes, now retired, was formerly a Keeper of Mineralogy at the Natural History Museum, London. He has a long research interest in the minerals and mineral deposits of Great Britain, including south and west England, the Channel Islands and Northern England, and is the author of numerous technical papers and popular publications including, with Peter Embrey, the book *Minerals of Cornwall and Devon*. Bob also has a particular interest in the history of collections and dealers. He is a past President of the Geologists' Association and the Russell Society. His research on minerals of the Mendip Hills is recognised in the naming of the new species, symesite, described from Merehead Quarry, Somerset in 2000.

Brian Young

Brian Young recently retired as the British Geological Survey's District Geologist for Northern England. He too has had a long interest in the geology and minerals of both the Lake District and northern Pennines, about which he has published numerous research papers and popular publications, including the *Glossary of Minerals of the Lake District*. Brian is a past President of the Yorkshire Geological Society and the Russell Society. He is currently an Honorary Research Fellow in the Department of Earth Sciences, at the University of Durham. His work on Northern England mineralisation is acknowledged in the new mineral, brianyoungite, found in the northern Pennines and described in 1993.

FACETED FLUORITE

Although too soft and brittle for use as a gemstone, clear fluorite crystals have occasionally been faceted. This fine selection of British fluorite includes many from Northern England. (E. A. Jobbins)

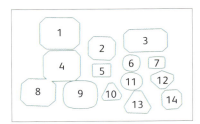

1. West Cumberland, 73.53 ct, pale bluish-green
2. Scordale, Westmorland, 25.24 ct, golden yellow
3. Weardale, Co. Durham, 47.74 ct, pale green
4. Beeralston, Devon, 47.15 ct, very pale bluish-green
5. Westmorland, 6.09 ct, greyish-mauve
6. Weardale, Co. Durham, 6.80 ct, pink
7. Weardale, Co. Durham, 5.54 ct, greyish-mauve
8. Rotherhope Fell mine, Cumbria, 35.79 ct, pinkish-mauve
9. Wheal Mary Ann, Cornwall, 33.70 ct, colourless
10. Rogerley mine, Co. Durham, 5.08 ct, green
11. Heights mine, Co. Durham, 9.01 ct, green
12. Cambokeels mine, Westgate, Co. Durham, 7.73 ct, green/mauve
13. Rogerley mine, Co. Durham, 11.36 ct, grey
14. Rogerley mine, Co. Durham, 9.73 ct, green

Introduction

NORTHERN England, as described in this book, extends from the Scottish border southwards to a line drawn rather arbitrarily across the country from Morecambe Bay eastwards to near Harrogate and thence northwards to the North Sea coast at the mouth of the River Tees. Included are the counties of Cumbria (comprising the former counties of Cumberland, Westmorland and parts of Lancashire), Northumberland, Tyne and Wear, Durham, and parts of Cleveland and North Yorkshire.

The region has long been justly famous for the great diversity of its mineral deposits and for the variety of minerals present within them. Indeed, apart from south-west England, no comparable part of Britain exhibits such an abundance of varied mineralisation.

The rocks which make up northern England, and which contain these minerals, also exhibit great variety. It is this varied geology which imparts to the region its scenic diversity. Much of the country is of high landscape value with three of England's National Parks – the Lake District, Northumberland and the Yorkshire Dales, – together with four Areas of Outstanding Natural Beauty (AONB), including the Northumberland Coast, Solway Coast, North Pennines, Forest of Bowland and Nidderdale. The North Pennines AONB is today designated as a Global Geopark.

At the centre of the Lake District is Scafell Pike, at 977 metres above sea-level, England's highest mountain. Surrounding Scafell are the distinctively craggy mountains (fells) which mark the outcrop of the Borrowdale Volcanic Group rocks. To the north are the smoother and less rugged fells formed of mudstones of the Skiddaw Group, the highest of which (931 m) is Skiddaw itself. North of this lies the rather rounded country of the Caldbeck Fells, which coincide closely with the outcrop of the Eycott Volcanic Group and associated igneous intrusions. The southern Lake District is characterised by undulating, locally rocky, country developed on the outcrop of the mudstones and sandstones now grouped together as the Windermere Super-Group. The eleven principal lakes occupy deep glaciated valleys in a crudely radial pattern centred about the highest fells of the Scafell massif.

The mineral wealth of the Lakeland fells has long attracted man, and as early as the 16th century the area was one of the world's principal copper producers. Many other minerals have been worked over the centuries, though the mines are silent and deserted today. Slate and hard rock quarrying are the district's only remaining extractive industries.

Writers and poets such as Southey, Ruskin and Wordsworth drew inspiration from Lakeland scenery, and the area provided Beatrix Potter with the home of Peter Rabbit, Squirrel Nutkin and Mrs Tiggywinkle. Such was the attraction of the Cumbrian landscape that, under the early guidance of Beatrix Potter, Canon Rawnsley and others, the National Trust, now one of the Lake District's biggest land owners, came into being.

To the south, west and north the Lake District fells give way rapidly to undulating coastal lowlands, though with fine red sandstone cliff scenery at St Bees Head. The Carboniferous limestones of west and south Cumbria contain large deposits of hematite which, fuelled initially with charcoal from the ancient woodlands of the southern Lake District and later with coal from the Cumbrian and Durham coalfields, provided the basis of the iron and steel-based heavy industries of Workington, Whitehaven, Millom and Barrow-in-Furness.

Beyond the gentle lowlands bordering the salt-marshes and sandbanks of the Solway coast and the border city of Carlisle, all underlain by Permo-Triassic rocks, the remote fells of the Bewcastle and Liddesdale areas along the Scottish border mark the outcrop of Lower Carboniferous sandstones, mudstones and limestones. From beneath these rocks in north Northumberland crop out the Devonian lavas of the Cheviot volcano.

Inland from the largely unspoilt Northumberland coast, with its wide sandy bays and spectacular castles, run a series of ridges formed of relatively resistant rocks such as the Fell Sandstone. One of the most prominent ridges, that formed by the intrusive Whin Sill, provides the superb naturally defensive site for the Roman frontier and World Heritage Site of Hadrian's Wall for a large part of its 70-mile course across the narrowest part of England.

South of Hadrian's Wall are the flat-topped fells of the northern Pennines with their extensive grouse moors. These rise to a spectacular west-facing escarpment which reaches 893m at Cross Fell, the highest point in the Pennines, overlooking the broad fertile lowland of the Vale of Eden between the Pennines and the Lake District. A distinctive feature of the northern Pennine landscape is the conspicuous terrace featuring of the hillsides which mark the outcrops of resistant beds of sandstone and limestone within the Carboniferous succession. Limestones are most prominent along parts of the Pennine escarpment and in the Wensleydale and Wharfedale areas. It is in this often remote upland country that the rivers South Tyne, Wear, Tees, Swale, Ure and Wharfe rise. Their valleys, or dales as they are commonly known in northern England, which drain eastwards to the North Sea, expose innumerable veins of lead ore which have, perhaps since Roman times or even earlier, been the basis for a large part of Britain's non-ferrous mining industry. Many of the area's small towns and villages such as Alston, Nenthead, Allenheads, Garrigill, Middleton-in-Teesdale, Reeth and Grassington, owe their origins or a significant part of their development to the once prosperous mining industry. Whereas ores of lead, iron and perhaps silver, were the minerals originally sought, the working of barite, witherite and fluorite were to be the mainstay of the mining industry for much of the 20th century. Although all mining has now ended, the appearance of the landscape still owes much to centuries of mineral extraction. Because of the local abundance of lead ore high in the Pennine hills, settlements were commonly established at much higher altitudes than elsewhere in Britain. In addition, as it was common for miners also to run small farms to supplement their food supply, the Pennine dales are characterised by a pattern of enclosed, 'intake', fields extending high up the valley sides.

Near the east coast the Lower Carboniferous rocks of the Pennine uplands are overlain by the Coal Measures of the Northumberland and Durham coalfield. One of Britain's oldest worked coalfields, the area is rich in remains and sites of industrial archaeological interest. Coal and local iron ores spawned heavy industries including ship building, which were once so important in towns such as Newcastle and Sunderland on the rivers Tyne and Wear. Coal mining has declined greatly in importance in recent years, and with the closure of the last underground mine in 2005, coal working is today restricted to a handful of opencast operations.

Within the coalfield lies the beautiful ancient cathedral city of Durham, one of Britain's few World Heritage Sites, built on the magnificent incised meander of the River Wear. To the east of Durham the coalfield is concealed beneath the Permian Magnesian Limestone, the outcrop of which is marked by a distinct belt of arable limestone country cut by deep gorge-like valleys (or denes) which run down to the pale yellow limestone cliffs of the Durham coast.

In a work of this sort it is impossible to do more than review the most important aspects of the geology. The very brief geological history which follows is intended to serve only as an introduction to the understanding of the region's mineral deposits. For the reader keen to explore the geology in more detail the following general accounts are recommended:

1. B. J. Taylor (*et al.*) (1971): *British Regional Geology. Northern England* (4th edition) (London: HMSO, for Institute of Geological Sciences).

2. N. Aitkenhead (*et al.*) (2002): *British Regional Geology – Pennines and adjacent areas* (British Geological Survey).

3. D. H. Rayner and J. E. Hemingway (ed.) (1974): *The geology and mineral resources of Yorkshire* (Yorkshire Geological Society).

4. F. Moseley (ed.) (1978): *The geology of the Lake District* (1st edition) (Yorkshire Geological Society).

5. G. A. L. Johnson (*et al.*) (1995): *Robson's geology of North East England* (Natural History Society of Northumbria).

A very large geological literature exists for the region; most of the more important references are given later in the Selected Bibliography.

BORROWDALE, CUMBERLAND

T. Allom – W. Le Petit
Dated *c.*1830

The distinctive craggy nature of the Borrowdale Volcanic Group rocks is clearly seen in this print. (©NHMPL)

Geological Background

IN order to appreciate northern England's minerals and mineral deposits in their true context, it is important to understand something of the area's geological diversity. The rocks that host these minerals record approximately half a billion years of geological history. This chapter reviews very briefly the varied earth processes and events that shaped these rocks and comments on some of the ways these rocks have been used over the centuries.

The oldest rocks exposed in the area date from the early part of the Ordovician period (approximately 500 million years ago). At this time the configuration of the world's oceans and continents was quite unlike that of today. The area which was eventually to become northern England lay in the southern hemisphere south of latitude 60°S on the margin of a continental mass known as Gondwana. A wide deep ocean, the Iapetus Ocean, separated this from the Laurentian continent, on the southern margin of which, a little south of the Equator, was the area that eventually became Scotland. During the Ordovician, as today, the earth's continental masses or plates were not static but were constantly moving relative to one another. At this time Laurentia and Gondwana were moving inexorably closer, with the Iapetus Ocean becoming progressively narrower.

Within this ocean, and the surrounding shelf seas, accumulated many hundreds of metres of mud, silt and sand which were to become the mudstones and sandstones of the Skiddaw Group of the Lake District and northern Pennines, and the similar rocks seen today as the Ingleton Group of the Yorkshire Pennines. Marine life was dominated by shoals of graptolites, although with some bottom-dwelling brachiopods and trilobites. Earth movements, related to the closure of Iapetus, produced deformation of these sedimentary rocks with some erosion before the onset of volcanic activity during the Middle Ordovician.

Subduction associated with the Iapetus closure established sub-aerial volcanic island arc complexes in which were formed the thick sequence of lavas and volcanic sediments of the Lake District: the Eycott Volcanic Group in the northern Lake District, and

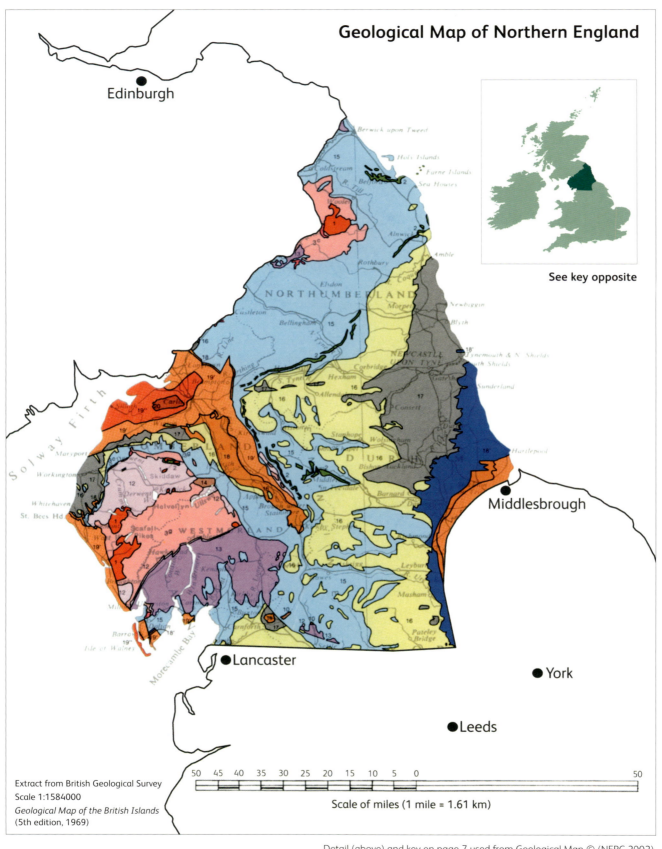

Geological Map of Northern England

Edinburgh

Berwick upon Tweed

Holy Islands

Farne Islands

Sea Houses

Coldstream

R. Till

NORTHUMBERLAND

Alnwick

Rothbury

Amble

Elsdon

Castleton

Bellingham

Morpeth

Newbiggin

Blyth

NEWCASTLE UPON TYNE

Tynemouth & N. Shields

South Shields

Corbridge

Hexham

Sunderland

Allend

Consett

Maryport

Skiddaw

CUMBERLAND

Derwent

Helvellyn

Ullsw

DURHAM

Stanhope

Wolsingham

Bishop Auckland

Hartlepool

Workington

Whitehaven

St. Bees Hd.

Scafell

Pikes

WESTMORLAND

Middleton in Teesdale

Barnard

Middlesbrough

Hawk

Ken

Sta

Sto

Ste

Barrow

Dalton

Leyburn

Carnforth

Masham

Morecambe Bay

Pateley Bridge

Isle of Walney

See key opposite

Solway Firth

Lancaster

York

Leeds

Scale of miles (1 mile = 1.61 km)

50 45 40 35 30 25 20 15 10 5 0   50

Extract from British Geological Survey
Scale 1:1584000
Geological Map of the British Islands
(5th edition, 1969)

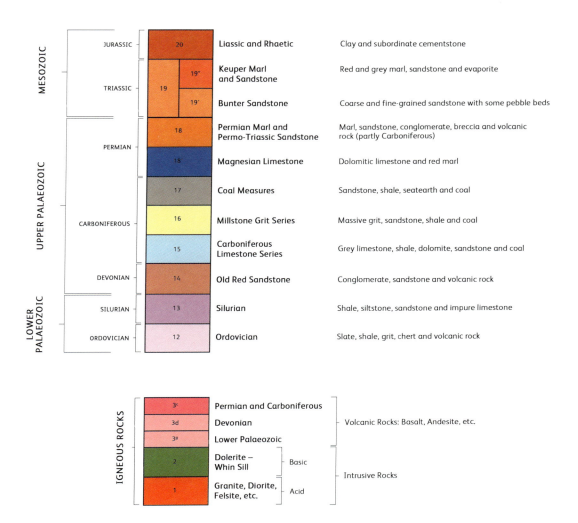

MESOZOIC	JURASSIC		20	Liassic and Rhaetic	Clay and subordinate cementstone
	TRIASSIC	19	19″	Keuper Marl and Sandstone	Red and grey marl, sandstone and evaporite
			19′	Bunter Sandstone	Coarse and fine-grained sandstone with some pebble beds
UPPER PALAEOZOIC	PERMIAN		18	Permian Marl and Permo-Triassic Sandstone	Marl, sandstone, conglomerate, breccia and volcanic rock (partly Carboniferous)
			18′	Magnesian Limestone	Dolomitic limestone and red marl
	CARBONIFEROUS		17	Coal Measures	Sandstone, shale, seatearth and coal
			16	Millstone Grit Series	Massive grit, sandstone, shale and coal
			15	Carboniferous Limestone Series	Grey limestone, shale, dolomite, sandstone and coal
LOWER PALAEOZOIC	DEVONIAN		14	Old Red Sandstone	Conglomerate, sandstone and volcanic rock
	SILURIAN		13	Silurian	Shale, siltstone, sandstone and impure limestone
	ORDOVICIAN		12	Ordovician	Slate, shale, grit, chert and volcanic rock

IGNEOUS ROCKS	3c	Permian and Carboniferous		Volcanic Rocks: Basalt, Andesite, etc.
	3d	Devonian		
	3ᵃ	Lower Palaeozoic		
	2	Dolerite – Whin Sill	Basic	Intrusive Rocks
	1	Granite, Diorite, Felsite, etc.	Acid	

the Borrowdale Volcanic Group in the central Lake District.

Rocks of the Eycott Volcanic Group form the northern hills of Binsey, the Caldbeck Fells and Eycott Hill. Although exposure is poor due mainly to an extensive mantle of glacial drift, this sequence is known to comprise a lower part in which andesite and dacite lavas and sills are interbedded with tuffs and volcanic breccias. A distinctive variety of basaltic andesite within this sequence contains striking phenocrysts of labradorite up to 20 mm long. The upper part of the Eycott Volcanic Group consists mainly of intermediate and acid tuffs.

Borrowdale Volcanic Group (BVG) rocks form the spectacular craggy fells of the central Lake District including the Scafell range, Great Gable, Langdale Pikes, the Coniston Fells and Helvellyn. BVG rocks also crop out in a narrow belt along the foot of the Pennine escarpment, and in a small inlier in Teesdale. Basalt, andesite and dacite lavas, with thick intercalations of volcanic sandstones and siltstones, comprise much of the Lower BVG sequence. Andesite sills are common locally.

The Upper BVG comprises a sequence of andesite and dacite lavas with thick rhyolitic ignimbrite sheets and, as in the Lower BVG, an abundance of interbedded volcaniclastic sedimentary rocks. Many of these Upper BVG rocks record the development and eventual collapse of a series of caldera. Some of the tuffs within the Upper BVG are famous for the presence within them of beautifully preserved accretionary lapilli. These are spherical accumulations of fine-grained volcanic dust accreted by raindrops falling through ash clouds. The resultant lapilli, when seen in section on broken or cut surfaces, suggest the local name 'birds eye' tuff for these rocks. Several of the andesites and dacites of the Lower BVG are well known for the occurrence within them of small euhedral almandine-pyrope garnets. These may be up to 4 mm across: good specimens have been obtained from a number of localities including Stockley Bridge at the head of Borrowdale and Lingmell Gill in Wasdale.

Man has long valued the special properties of some of these volcanic rocks. A bed of fine-grained siliceous tuff within the Upper BVG provided the raw material from which Neolithic man fashioned beautifully polished stone axes. Large numbers have been found over a wide area of Britain, and damaged and incomplete examples may be found occasionally at the sites of the Neolithic axe 'factories' in the Great Langdale area.

Cessation of volcanic activity in the late Ordovician was followed by more earth movements and a resumption of marine sedimentation in the by now much narrower Iapetus Ocean. The continual plate movement responsible for this progressive closure by now had brought the area which was to become northern England farther north towards the Equator.

Late Ordovician and Silurian sedimentary rocks, collectively known as the Windermere Super-Group, form much of the southern Lake District and crop out in the Malham and Ribblesdale areas of the Yorkshire Dales. Marine limestones at the base contain brachiopods, corals and trilobites. These are followed upwards by a thick succession of deep water mudstones, siltstones and sandstones, many of which formed as turbidites. Two episodes of acid volcanism in the lower part of the Windermere Super-Group gave rise to local deposits of ignimbrite and volcaniclastic rocks. The siltstones, mudstones and sandstones of the uppermost part of the Windermere Super-Group indicate a return to relatively shallow marine conditions. These were deposited in a foreland basin which occupied the position of the former Iapetus Ocean when this was destroyed by the final collision of the Laurentian and Gondwanan plates about 410 million years ago.

The Caledonian earth movements which followed this collision caused major folding and faulting of earlier formed rocks and imparted a pronounced and widespread cleavage to many of the rocks. This cleavage was especially well-developed in finer-grained sedimentary rocks, many of which, including those within the Borrowdale Volcanic Group and parts of the Windermere Super-Group, were transformed into slates. The 'Lakeland green slates', formed from sedimentary units within the BVG, have long been worked for roofing and ornamental stone from quarries at Honister, Borrowdale, Kirkstone, Coniston, Tilberthwaite and Elterwater. Magnificent sedimentary structures are a feature of many of these slates which are still quarried at Kirkstone and Coniston.

THRANG CRAG SLATE QUARRY, GREAT LANGDALE

T. Allom – W. Le Petit (published by Fisher, Son & Co., 1834)

The cavernous interior of one of the Lake District's underground slate quarries, in which cleaved volcanic sediments of the Ordovician Borrowdale Volcanic Group were worked, is spectacularly shown in this print. (©NHMPL)

Accompanying these movements was an important phase of igneous activity. Intrusion of large volumes of igneous rock beneath the Lake District rocks began during the Late Ordovician with the formation of the Carrock Fell and Haweswater complexes; each were composed mainly of basic rocks such as gabbros, but with subordinate microdiorites and microgranites. Intrusion of the Carrock Microgranite has been dated at 452 ± 4 million years. Parts of the Carrock Gabbro are rich in magnetite and ilmenite, which may locally comprise as much as 30% of the volume of the rock. Good examples of this ilmenite-rich facies may be seen at Furthergill Sike, east of Carrock Fell.

Emplacement of granitic intrusions followed. The earliest of these, about 450 million years ago, were the Threlkeld, Eskdale and Ennerdale intrusions in the Lake District. A little later, about 390 million years ago, the granites of Shap, Skiddaw and the concealed granites of Weardale and Wensleydale beneath the Pennines, were emplaced. Geophysical studies have revealed that much of the Lake District is underlain by a large composite granite batholith of which the granites seen today at outcrop are constituent parts. Recent reinterpretation of geophysical data suggests that the granites beneath the Pennines may be similar composite batholiths, of which the Weardale and Wensleydale granites form part.

Despite sharing a common source, each of these granites is petrologically distinctive, although with some variation in rock types within each intrusion. Probably best known, and most distinctive, of the Lake District granites is that of Shap. Much of this comparatively small intrusion consists of a strikingly porphyritic rock in which abundant phenocrysts of pink orthoclase perthite up to 40mm long, some of which exhibit striking rapakivi texture, occur in a coarse-grained matrix of quartz, orthoclase, oligoclase and biotite. The rock provides an excellent building and ornamental stone and has long been quarried from the Shap 'Pink' Quarry. Beautiful polished slabs of Shap Granite may be seen in buildings throughout Britain. Parts of the Threlkeld and Eskdale intrusions have been quarried mainly for local use as building stone or crushed rock aggregate. A biotite-rich granodiorite in the southern part of the Eskdale intrusion has been worked for building stone at Broad Oak quarries near Wabberthwaite.

Much of the northernmost outcrop of the poorly exposed Skiddaw Granite in Grainsgill consists of quartz-mica greisen. This rock was produced by the metasomatic effects of high temperature fluids shortly after the emplacement of the granite. These same fluids were responsible in part for the tungsten-bearing quartz veins which cut the greisens and associated rocks, and

LEFT: SHAP GRANITE QUARRY

Also known as the Shap 'Pink' Quarry. (©NHMPL)

ABOVE: SHAP GRANITE

Polished slab of Shap Granite showing distinctive orthoclase perthite phenocrysts up to about 2 cms long, and a prominent black rounded xenolith. (B. Young)

which were worked at Carrock Fell mine. Similar greisens, some which contain abundant topaz, are common adjacent to the contacts of the Eskdale Granite.

The Dufton Microgranite, a small intrusion within the Lower Palaeozoic rocks of the Cross Fell Inlier at the foot of the Pennine escarpment, is well known for its euhedral muscovite phenocrysts, up to about 10 mm across.

The intrusion of these large bodies of igneous rock produced considerable alteration of the surrounding rocks. A well developed thermal metamorphic aureole within the Skiddaw Group rocks surrounds the Skiddaw Granite. Hornfelsed mudstones closest to the granite contact contain biotite and abundant cordierite: further away from the contact the hornfels contains conspicuous blade-like crystals of the chiastolite variety of andalusite. Extensive metamorphism and metasomatism of the BVG rocks across much of the Lake District has produced such minerals as biotite, hornblende and abundant pale green epidote. Notable localities for specimens of the latter include Walla Crag in Borrowdale. The metasomatic alteration of BVG rocks around the Shap Granite has given rise to a large suite of minerals best seen in the Shap 'Blue' Quarry which is worked as a source of roadstone and aggregate. Conspicuous here are bright green epidote, dark green actinolitic hornblende and coarse-grained dark reddish brown grossular-andradite garnet, together with smaller amounts of magnetite, pectolite, laumontite and, rarely, a little prehnite. One of the earliest major periods of metalliferous mineralisation within the Lake District resulted from mobilisation of hydrothermal fluids during this period of earth movement and igneous emplacement.

An unusual application of the properties of the cordierite hornfels from the Skiddaw aureole was in the making of stone 'xylophones', or more correctly 'lithophones', two of which are to be seen in Keswick Museum. The musical qualities of some Skiddaw Group rocks were first noted by Peter Crosthwaite, of Crosthwaite near Keswick, in his diary of 1785. One of these instruments, still in playable condition, in Keswick Museum was built in the early 19th century by Joseph Richardson, one of a family of Keswick stone masons. The stone keys, which span a range of seven octaves, are all of cordierite hornfels and are said to have taken 13 years to collect and shape. Many concerts were given in Britain and across Europe, and in 1848, by command of Queen Victoria, the instrument was played at Buckingham Palace. Parts of a much more modest 'lithophone', with keys of cordierite hornfels, are also in Keswick Museum. Other such instruments include those in the Kendal Museum and the Ruskin Museum at Coniston.

The enormous geographical changes brought about by the Caledonian Orogeny provided the setting for the next major episode of earth history – the Devonian period. At the beginning of this period much of what was to become Britain was land, probably at a latitude of about 25°S. Desert uplands or mountains occupied the area of the Southern Uplands, Lake District and Pennines. As northern England has few rocks of Devonian age, the history of the region during this period is somewhat uncertain. However, the Cheviot area was occupied by a volcano which erupted mainly andesitic lavas and ashes. The lavas locally contain beautiful agates. Towards the close of this volcanic episode the lavas were intruded by large masses of granite, parts of which are exposed today at the centre of the Cheviot Hills, including Cheviot itself. A small associated body of bright red microgranite, near Alwinton on the southern side of the Cheviot Hills, is quarried for roadstone which is used to surface the Mall in London, and other roads where its distinctive red colour is valued.

Erosion of the precursors of the Lakeland mountains and Pennines piled up fans of rock debris seen today as the Mell Fell Conglomerate and the Polygenetic Conglomerate of the Pennine escarpment. No other deposits of Devonian age are known at outcrop in northern England, although red sandstones and siltstones, known collectively as the Old Red Sandstone, are present immediately north of the Scottish border in the Jedburgh area.

The Carboniferous period, during which most of the region's rocks were deposited, saw a progressive submergence of the Devonian desert land surface beneath a tropical sea: the area which was to become northern England now lay just a little south of the equator. Submergence was not uniform across the region, but was controlled by a number of important structural features. At this time a high ridge extended

A selection of polished agates from the Cheviot Hills, Northumberland. Found in sites in weathered lavas at Buckhams Wall Burn, in shingle of the River Coquet near Holystone, and in soil gravels of the Mindrum area. The largest specimen is about 5 cm across. (©NHMPL)

The larger of the stone 'xylophones' housed in Keswick Museum. (©NHMPL)

eastwards from the Isle of Man across the area of the Lake District to the northern Pennines. To the north lay another upland landmass in the area of the Southern Uplands. As the Lower Carboniferous sea advanced, these upland or 'block' areas subsided less rapidly than the surrounding 'troughs' or 'basins', due to the buoyancy given to the 'blocks' by the presence within them of large bodies of granite. In consequence, the sequences of strata deposited in the troughs are much thicker than the equivalent sequences deposited on the blocks. The boundaries between troughs and blocks were comparatively narrow hinge zones along contemporary fault lines which were initiated during the Caledonian earth movements. This differential subsidence, which was most pronounced during Lower Carboniferous times, had virtually ceased by Upper Carboniferous times when the Coal Measures were deposited. Despite these marked variations in subsidence rates, and consequent thicknesses of deposited sediment, sedimentation everywhere almost always kept pace with subsidence. The depositional surface at any time across both blocks and troughs was almost horizontal, with limestones and marine mudstones deposited in maximum water depths of a few tens of metres.

Some of the region's earliest Carboniferous rocks

are conglomerates composed of debris derived from the pre-Carboniferous land surface. Examples of these may be seen in the Cockermouth area, on the Pennine escarpment and in Upper Teesdale, and locally on the eastern flanks of the Cheviot Hills. Local episodes of volcanic activity, probably related to movement along major faults, erupted a few flows of basalt lavas in the Border area and around Cockermouth.

Submergence beneath the warm tropical Lower Carboniferous sea resulted in the deposition of thick sequences of limestone initially in south Cumbria and the Shap and Craven areas, and eventually over much of the region. In the Northumberland-Solway 'Trough' the Lower Carboniferous rocks comprise a thick sequence of mudstones, sandstones and limestones. These record a complex interplay of marine and freshwater conditions as a large delta, or series of deltas, spread south-westwards across the region bringing sediment from the uplands which lay in the area of what is now southern and central Scotland and parts of the North Sea.

Some of the region's Lower Carboniferous limestones are richly fossiliferous with abundant corals, brachiopods and crinoids. A few of these rocks have been used as polished ornamental stones. Best known of the Carboniferous ornamental stones is the so-called 'Frosterley Marble'. This is not a true marble, but is a dark grey bituminous limestone crowded with large, beautifully preserved white fossils of the solitary coral *Dibunophyllum bipartitum*. It occurs as a widespread bed, up to 0.75 metres thick, within the Great Limestone of Weardale and Teesdale. It takes an excellent polish and has been used as an ornamental stone since at least the 14th century in many churches in the region and beyond. Especially fine examples are to be seen in Durham Cathedral. Some of the pale fawn, coarsely crinoidal facies of the Undersett Limestone near Scotch Corner is still used as a polished facing stone, known in the trade as Swaledale Marble.

The continued submergence of the Lake District and Pennine Blocks was accompanied by a progressive southwards extension of the delta complexes from the north. Over much of the region Lower Carboniferous sedimentation was characterised by an alternation of marine and deltaic influences. The resulting cyclical repetition of lithologies, limestone, mudstone, sandstone and locally coal, may be explained by alternating periods of relatively slow and rapid subsidence. During the former periods, delta lobes advanced across the area from the north depositing mud and sand. Occasionally deltas were built up to above sea level, so allowing the formation of swamp forests, now preserved as coal seams. More rapid subsidence allowed warm clearer water from the south to overwhelm the deltas and led to the deposition of marine limestones. The sequence of rocks deposited during each such cycle of events is termed a 'cyclothem'. The distinctive Lower Carboniferous cyclothemic sequences of the Pennines are termed the 'Yoredale' facies after an early name 'Yoredale', for Wensleydale, where they are especially well developed. Differential erosion of these rocks gives rise to the distinctive terraced hillsides of the Pennine valleys.

By the close of Lower Carboniferous times, deltaic influences were dominant across the region. It was this style of deposition that was to characterise the remainder of Carboniferous time. Over large parts of the region the Upper Carboniferous rocks consist of a rhythmic succession of mudstone, sandstone and locally limestone similar to the Yoredale cycles of the Lower Carboniferous, although with limestones generally much thinner and locally absent. Non-marine rocks, locally including coal seams, occupy a much greater proportion of each cycle. Thick, coarse-grained, gritty, cross-bedded sandstones are common. These are often laterally impersistent and many have been shown to occupy distributary channels in the Carboniferous deltaic plain.

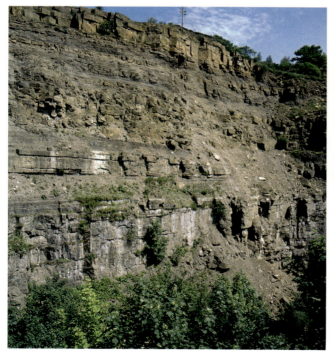

LEFT: ROGERLEY QUARRY, WEARDALE

Carboniferous mudstones and sandstones, showing channelling features, overlie the Great Limestone that forms the lower part of the face, above the trees. (©NHMPL)

ABOVE: 'FROSTERLEY MARBLE'

A distinctive bed of dark grey limestone within the Great Limestone of the northern Pennines, distinguished by an abundance of the solitary coral *Dibunophyllum bipartitum*. Although not a true marble, the rock has been much used as an ornamental stone. (©NERC)

Upper Carboniferous mudstones and sandstones overlie the limestones of the Lake District margin. They also form a more or less continuous outcrop along the eastern margin of the Lower Carboniferous rocks of the Pennines, with outliers forming caps to some of the higher limestone hills to the west, such as Cross Fell, Ingleborough, Whernside and Pen-y-ghent. Many of the sandstones are ideally suited to the making of millstones and grinding stones, and it is this former widespread use which gave the name 'Millstone Grit' to this division of the Upper Carboniferous of the central and southern Pennines. Gritstones are best developed in the Yorkshire Pennines where an unconformity, which separates them from the underlying limestones, marks a period of Carboniferous folding and erosion.

The final chapter in the Carboniferous history of northern England saw the deposition of the rocks which were to give to geology the term 'Carboniferous'. This was the period when, across much of what was to become Britain and Europe and much of North America, the great thickness of coal-bearing rocks, the Coal Measures, were laid down.

During Coal Measures times the deltaic plains established earlier in the Carboniferous were frequently colonised by luxuriant swamp forests composed of giant relatives of today's ferns, club mosses and horsetails, together with extinct ancestors of modern conifers. A continuous supply of plant debris onto the forest floor accumulated in the stagnant waterlogged conditions, forming thick layers of peat. Periodic influxes of sediment, resulting from local flooding, caused by the breaking of river banks or rejuvenation of sources of sediment supply, overwhelmed the forests and peat beds, burying them beneath layers of mud or sand to form coal seams, which may be up to 4 m thick but are usually thinner. On infrequent occasions widespread invasion of the deltaic plains by marine water deposited mudstone with distinctive marine faunas. These 'marine bands' are extremely important in enabling the detailed interpretation and correlation of the succession of rocks within and between coalfields. The cyclical pattern of sediment accumulation, forest growth and subsequent overwhelming by more sediment, was repeated many times during Coal Measures times. Compaction and lithification of the sediments produced the succession of mudstones, sandstones and coal seams seen today.

Clear evidence of contemporary conditions may be found in these rocks. Coal seams typically rest on a bed of fossilised soil, known as a palaeosol or 'seatearth', which may be either a sandstone or a mudstone in which fossil rootlets from the overlying coal swamp are abundant. Whereas most of these fossilised roots are small and fragmentary, large complete tree stumps and root systems are occasionally found in growth position. A fine example from a nearby quarry is preserved in Stanhope churchyard in Weardale. Under the prevailing conditions of tropical weathering, seatearths are typically intensely weathered and leached of a number of chemical elements. Mudstone seatearths are typically highly aluminous clays which, because they can be used for making refractory products such as fire-bricks or furnace linings, are commonly known as fireclays. The term 'seggar' is often used in northern England for fireclay. Sandstone seatearths are commonly hard, very silica-rich rocks known as 'ganister'. Like fireclays these too are well-suited for making refractory products such as furnace linings.

Microscopic examination of the coal commonly reveals remains of plant tissue. More easily recognisable plant remains, including beautifully preserved leaves, are found locally in mudstone partings within the seams or in the overlying measures. Fossils of fresh-

AIR SHAFT AT WALLSEND QUARRY
Thomas Hair, engraving, 1839

Hair was a London artist who, between 1830 and 1839, painted a number of watercolours of collieries in the Northumberland and Durham coalfield. Many of these were made into engravings which were published in 1839.

LEFT: HIGH FORCE, TEESDALE

The River Tees here plunges over hard, columnar jointed dolerite of the Whin Sill, at England's biggest waterfall. The Carboniferous Tynebottom Limestone crops out at the foot of the fall. (B. Young)

ABOVE: APOPHYLLITE CRYSTALS

Pale brown apophyllite crystals on radiating pale buff pectolite, from a vein in the Whin Sill at Copt Hill Quarry, Cowshill, Weardale. The specimen is 17 cm across. It was presented to the Natural History Museum by Messrs Monkhouse and Peart, operators of the quarry, in March 1911. (NHM spec. no. BM 1911, 168; photo. ©NHMPL)

water bivalves, 'mussels', are also commonly found in these beds, and more rarely the remains of amphibians which lived in the coal-forming swamps. Marine fossils such as goniatites, brachiopods and crinoids occur in the marine bands.

The area being described in this book includes two of Britain's major coalfields: the Northumberland and Durham coalfield, and the Cumbrian coalfield. Smaller, isolated areas of Coal Measures rocks have been worked in the Tyne Valley, Stainmore and Ingleton areas. The history of coal mining extends back over several centuries and both fields have been of prime importance in the development of coal mining technology.

Throughout northern England, Carboniferous times were brought to an end by the folding and faulting of the Hercynian earth movements. Accompanying these movements was the intrusion of a large volume of basaltic magma within the Carboniferous rocks of Northumberland and the northern Pennines, as a suite of more or less horizontal sheets or sills and associated dykes, known collectively as the Great Whin Sill. This underlies much of south eastern Northumberland, the northern Pennines and the Durham coalfield. It crops out on the Northumberland coast in the Farne Islands,

at Bamburgh and Dunstanburgh castles, and forms the spectacular crags along which Hadrian's Wall is built. In the northern Pennines it gives rise to the waterfalls of High Force, Low Force and Cauldron Snout in Teesdale, and the amphitheatre-like cliffs of High Cup Nick on the Pennine escarpment. Whereas over much of this huge area the intrusion occurs as a single sill, in places it splits into two or more separate leaves. Recent research has shown that the Whin Sill actually consists of a group of closely related intrusions. All consist of fine to medium-grained quartz dolerite, although irregular coarse-grained bodies of dolerite pegmatite are present in places, especially in the northern Pennines, where the Great Whin Sill reaches its maximum thickness of over 70 m. Joints within the sill in the northern Pennines commonly contain pectolite and calcite, although in places analcite, apophyllite, chabazite, stilbite and stevensite have also been found. Contact metamorphism of the adjacent shales and impure limestones has produced calc-silicate rocks containing garnet, vesuvianite and locally magnetite: the purer Melmerby Scar Limestone in Upper Teesdale has been converted into a coarse-grained marble. This rock, known from its characteristic weathering as 'sugar limestone', supports an important relic arctic alpine flora, including the spring gentian (*Gentiana verna*).

The Whin Sill is generally regarded as the original sill of geological science. 'Sill' was a northern England miner's term for any horizontal bed or body of rock. 'Whin' was the quarrymen's name for a hard rock, and is supposed to have been derived from the sharp whistling noise made by fragments flying off a block when broken.

Uplift and erosion in Permian times converted the region into a desert. Bare, rugged hills with abundant screes occupied the area of the present day Lake District and Pennine escarpment. Angular debris, washed by flash floods into the adjacent lowlands, accumulated to form the breccias or 'brockrams' seen today in west Cumbria and the Vale of Eden. Extensive dunes of coarse sand in the area destined to become the Vale of Eden are seen today as the Penrith Sandstone, one of northern England's best known building stones. Similar dunes on the more gently sloping land surface east of the Pennines, and extending into what is today the North Sea, became the Yellow Sands, which today in County Durham are an important source of building sand. They also, in parts of the North Sea, form an important hydrocarbon reservoir.

Late Permian times saw the inundation of the eastern part of the region beneath the advancing waters of the Zechstein Sea, the earliest deposit of which is the Marl Slate. This comprises up to about 2 m of carbonaceous calcareous mudstone in which complete or almost complete fish fossils are locally common. The bed is also noteworthy for the presence within it of comparatively high concentrations of lead, zinc and copper. The Marl Slate is the English correlative of the Kupferschiefer of Germany and eastern Europe where copper concentrations are sufficiently high for the rock to have been worked as an ore.

A complex succession of tropical marine limestones and dolomites, collectively termed the Magnesian Limestone, overlies the Marl Slate in north east England. Included in these rocks are richly fossiliferous reef limestones, oolitic limestones and spectacular collapse breccias, indicative of the removal by dissolution of thick beds of gypsum and anhydrite. The equivalents of these occur at depth in the Hartlepool and Billingham area, where they were formerly mined, and offshore

in the North Sea. A remarkable limestone, in which removal of dolomite – dedolomitisation – has caused extensive recrystallisation, is the concretionary limestone of the Sunderland area. Large parts of this formation consist of spherical or elongate, occasionally fan-shaped, calcite concretions in a rather softer, more dolomitic, matrix. Spectacular rounded concretions, up to 150 mm across, have earned the rock the local name of 'cannon-ball limestone'.

Inland, the Magnesian Limestone outcrop forms a prominent west facing scarp, scarred with numerous quarries for limestone and the underlying Yellow Sands, overlooking the Coal Measures outcrop. At the coast it forms an almost continuous line of pale yellow sea cliffs.

In north and east Yorkshire the Permian marine limestones pass upwards into a series of calcareous mudstones or marls deposited in a near-shore or coastal plain environment. Beds of gypsum and anhydrite thicken eastwards and are overlain by beds of halite and potash near the present coast and beneath the North Sea. The thick beds of salt and anhydrite in the Hartlepool and Billingham areas formed the basis of the huge chemical industry of Teesside. Anhydrite is no longer mined here, but salt is still extracted by solution mining. Potash and rock salt are mined today from Britain's deepest mine, Boulby mine, near Staithes on the Yorkshire coast, a short distance south of the area covered by this book.

Near the Cumbrian coast a few metres of dolomitic limestones, believed to correlate with the Magnesian Limestone of County Durham and Yorkshire, were deposited close to the margins of a tropical sea which occupied part of the area now occupied by the present Irish Sea, and which may on occasions have connected with the Zechstein Sea. Much of the Permian succession of Cumbria, however, consists of a sequence of mudstones and siltstones, known as the St Bees Shales and Eden Shales, deposited on or near to a coastal plain on which periodic evaporation produced beds of gypsum and anhydrite. These evaporite deposits have been worked in the Whitehaven area, and gypsum is still mined in the Vale of Eden around Kirkby Thore for the manufacture of plaster products. Some alabaster for ornamental use was mined in the 19th century at Barrowmouth, near Whitehaven. A very distinctive variety of porphyroblastic gypsum, in which stellate clusters of brown gypsum crystals occur within a fine-grained alabastrine gypsum matrix, and known as 'daisy gypsum', occurs in the Kirkby Thore area.

The St Bees and Eden Shales pass upwards into a very thick succession of fine-grained red sandstones deposited in a fluviatite environment, interpreted as part of a huge braided river system. This sandstone sequence collectively forms part of the Sherwood Sandstone Group. In Cumbria the major part of this is known by the local name of St Bees Sandstone, after St Bees Head, south of Whitehaven, where it forms spectacular red coastal cliffs. The Permian-Triassic boundary is rather arbitrarily placed at the base of the St Bees Sandstone. The St Bees Sandstone provides an excellent freestone which has been worked at many quarries throughout Cumbria: it is still quarried near St Bees. This dull red stone is a characteristic feature of numerous Cumbrian buildings, notable examples of which include Carlisle Cathedral and Castle and Furness Abbey.

Non-marine red mudstones, belonging to the Mercia Mudstone Group, and known in north Cumbria as the Stanwix Shales, overlie the Sherwood Sandstone Group. Exposures of these rocks are rare, although boreholes in the Carlisle and Barrow-in-Furness areas reveal many hundreds of metres of red mudstones with rare sandstones and limestones and, in places, beds of gypsum and halite. Salt was produced during the 19th century by brine pumping from these beds in the Barrow-in-Furness area.

Apart from the widespread deposition of marine Liassic rocks, probably across much of the region, and represented today only by a small outlier of fossiliferous mudstones in the Carlisle area, there are few deposits to record the geological history of the region until the Quaternary.

Folding and faulting in Tertiary times was accompanied by the intrusion of a small number of basaltic dykes, including the Acklington Dyke of Northumberland and the Armathwaite-Cleveland Dyke of Cumbria and the northern Pennines. These are associated with the igneous activity then occurring in the west of Scotland and Northern Ireland and related to the opening of the North Atlantic.

Uplift of the whole region set the scene for the development of the present landscape, the basic form of which was much modified during the Quaternary glaciations. Evidence for the enormous erosive power of valley glaciers is apparent in the U-shaped valleys, hanging valleys and corries of the Lake District. The abundance of erratic boulders of distinctive rock types scattered across parts of the region attest to the widespread transportation of debris (glacial drift) by ice sheets and glaciers. Not only are boulders of north of England rocks common in the region's 'drift' deposits, but blocks of Scottish and Scandinavian rock types are found along parts of the Irish Sea and North Sea coasts respectively.

Incised river valleys, including the striking Wear gorge at Durham, raised beaches, submerged forests and river terraces, were formed during the post-glacial period as sea-level adjusted to its present level.

During the 19th century, deep excavations in the alluvium of the River Tyne at Jarrow Slake yielded curious crystals, then thought to be a new mineral that was named 'jarrowite'. These subsequently have been shown to be pseudomorphs after the unusual calcium hexahydrate mineral ikaite which is known to crystallise only at temperatures near to freezing point.

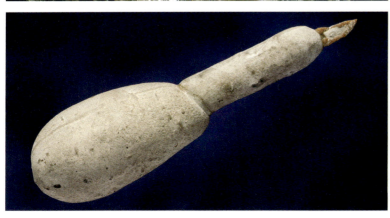

Mineralisation

NORTHERN England contains a remarkable wealth of mineralisation. Indeed, apart from south-west England, no other part of Britain exhibits such an abundance and diversity of metalliferous deposits. It has been estimated that about half of the mineral species known to occur in Britain are present in the Lake District alone.

It is impossible and inappropriate in a book of this sort to attempt detailed descriptions of individual deposits or groups of deposits. Excellent accounts of these are listed in the Bibliography. In the following pages the main groups of deposits are outlined in sufficient detail to provide a background against which the minerals, which form the main subject of the book, may be understood. The deposits reviewed here are all epigenetic: that is they have been introduced into the rocks in which they are found, and are therefore younger than the host rocks. Some of the region's minerals formed at the same time as their host rocks, i.e. syngenetic minerals, have been mentioned in the discussion of the geological background.

For convenience, the region's main mineral deposits will be considered in three main geographical areas: the Lake District, the northern Pennines, and west and south Cumbria.

The Lake District

Epigenetic mineralisation here is widespread within rocks of the Skiddaw, Eycott and Borrowdale Volcanic Groups, and major igneous intrusions: mineralisation is very rare within rocks of the Windermere Super-Group. Veins are commonly concentrated above ridges on the roof region of the concealed Lake District Batholith, or above its north and south walls. The several main suites of mineral deposits reflect different episodes of mineralisation which range in age from Late Ordovician to post-Triassic. Studies of intensely metasomatised Skiddaw Group rocks in the Crummock

OPPOSITE PAGE

Force Crag vein in No.1 Stope Force Crag mine, Braithwaite, Cumbria in 1991. The vein, here between Skiddaw Group mudstone wall-rocks, is unusually wide and contains an abundance of sphalerite with a conspicuous band of white barite near the right of the picture. The orange-clad figures are the late Mike Sutcliffe (left) and Lindsey Greenbank (right). Brian Young, in yellow overalls, leans against the hanging wall of the vein. (©NERC)

Water area suggest that these rocks were a likely source of metals in parts of the area.

Graphite Deposits

The graphite deposit at Seathwaite in Borrowdale is one of northern England's most famous mineral deposits and certainly has one of the longest histories of working. Graphite occurs here as very pure nodules, which range in size from one millimetre up to over one metre in diameter, in a series of at least eight steeply-inclined vein-like, or pipe-like, bodies within an altered dolerite intrusion of Ordovician age, within rocks of the Borrowdale Volcanic Group. Other introduced minerals are scarce, but include pyrite and chlorite.

The Seathwaite deposit is probably truly unique as, unlike most other graphite deposits in the world, the Seathwaite Graphite is not directly associated either with high grade thermal or regional metamorphism of a carbonaceous rock. A wholly satisfactory explanation of the deposit's origin has yet to be established, although a source of carbon with organic-rich Skiddaw Group mudstones at depth, with transportation of the carbon as carbon monoxide in the intruding dolerite, has been suggested.

First discovered and worked in the 16th century, the graphite was mined intermittently until 1891. Its main early use was in the making of moulds for casting cannon balls, although its supposed medicinal properties were also advocated. The marking of sheep with Borrowdale Graphite, claimed by some writers as its first use, anticipated by about two centuries its employment as the raw material for the Keswick pencil industry.

Graphite is also reported to have been worked in Bannerdale to the east of Blencathra. 'Bannerdale graphite' appears to have been a graphite-rich mudstone in the Skiddaw Group within the aureole of the Skiddaw Granite. Although the rock is very dark grey and capable of writing on paper, analyses suggest there is as little as 5% carbon within it. It is said to have been mined last century to make grate polish, although the amounts raised must have been extremely small.

Tungsten Veins

A suite of veins, worked until 1981 at Carrock Fell mine in the northern Lake District, comprises the only known economic concentration of tungsten mineralisation in Britain outside south-west England. The veins strike approximately N-S through the northernmost outcrop of Skiddaw Granite, here mostly altered to greisen, and the adjoining gabbros and hornfelsed slates. White quartz, much of it in large crystals, com-

LEFT: GRAPHITE NODULES

Typical specimen of graphite nodules up to 2 cm across in altered dolerite. Seathwaite mine, Borrowdale, Cumbria. (©NERC)

Raw graphite and graphite pencils made from Seathwaite graphite, from Keswick Museum. (©NHMPL)

ABOVE: RUSSELLITE

Russellite, from Buckbarrow Beck, Corney Fell, Cumbria. Pale yellowish green spherules up to 0.25 mm across on quartz. Russellite was one of a number of rare bismuth supergene minerals found in this small vein. (©NERC)

prises the major portion of the veins, with wolframite and scheelite locally conspicuous as large crystals. Arsenopyrite, pyrite, pyrrhotite and sphalerite are common, with smaller amounts of native bismuth, bismuthinite, bismuth sulphotellurides, molybdenite, antimony sulphosalts and rarely a little gold. Non-metallic minerals include muscovite, apatite, fluorite and calcite. Numerous other minerals, many of super-gene origin from the oxidised portions of the veins, have been recorded. A number are described and illus-trated in a companion volume which describes the minerals of the Caldbeck Fells. (See Cooper and Stanley [1990]).

Much research has been carried out on the Carrock Fell veins. The veins are believed to have been emplaced during a complex series of mineralising events in Lower Devonian times from fluids derived from the final stages of cooling of the Skiddaw Granite.

Although not present in economic quantities, tungsten mineralisation has also been recorded from the Shap Granite where small crystals of scheelite have been found in drusy cavities. A quartz vein which carries scheelite, ferberite and chalcopyrite occurs in the Eskdale Granodiorite near Bootle. The oxidised outcrop of this vein is notable for the local abundance within it of the very rare minerals russellite, bismuto-ferrite, eulytite, namibite and cuprotungstite.

Antimony Veins

A few tons of stibnite were mined during the 19th century at Robin Hood mine, near Bassenthwaite, where it occurred in quartz veins associated with a basic intrusion in Skiddaw Group mudstones. Apart from an open level, in which no mineralisation can today be seen, little remains to mark the site of this venture.

Antimony-bearing veins occur at a number of local-ities throughout the Lake District, although no other workable deposits have been found. Stibnite, cosalite and other antimony sulphosalts are present in parts of the Carrock Fell tungsten veins. Stibnite, berthierite, jamesonite and zinckenite are abundant in a vein at Wet Swine Gill, near Carrock Fell; and berthierite and jamesonite-bearing quartz veins may be seen at Hogget Gill, Patterdale. Stibnite has been reported from quartz veins on St Sunday Crag near Helvellyn, although a recent detailed survey of this area failed to locate any evidence for them. The age of Lake District antimony mineralisation is unknown, although that at Carrock Fell mine and Wet Swine Gill appears to be related to the Lower Devonian tungsten mineralisation.

An intriguing occurrence of stibnite is that at Trout-beck Station where, on two occasions in the 19th century, erratic boulders of pure stibnite, one weighing 50 kilograms, were found in boulder clay. The *in situ* source has never been located.

Copper Veins

Veins in which copper minerals form the major metal-lic component are one of the major suites of Lake District deposits. Many were formerly of great econ-omic importance and were the basis of the Elizabethan copper industry when the Lake District was one of the world's leading centres of copper mining. The main concentrations of this mineralisation are in Coniston, the Vale of Newlands, parts of the Caldbeck Fells, Ulpha, Black Combe and Haweswater.

Typically these are fissure veins which occupy faults, mostly in Borrowdale Volcanic, Eycott Volcanic or Skiddaw Group rocks. The main ore mineral is usually chalcopyrite accompanied in many instances by appreciable quantities of arsenopyrite and pyrite, in a gangue assemblage dominated by quartz, chlorite and dolomite and, in places, stilpnomelane. Minor consti-tuents include tennantite, pyrrhotite, native bismuth, bismuthinite, bismuth sulphoselenides and sulpho-tellurides, galena and sphalerite. Traces of gold are known locally. In the veins of the Coniston area, small local concentrations of the nickel and cobalt ores nickeline, rammelsbergite, safflorite and skutterudite, were mined alongside copper ores during the 19th century. Also in the Coniston area, some of the copper veins contain appreciable quantities of magnetite. This mineral was particularly common in the Bonsor vein where its increasing abundance with depth was one of the factors which led to the abandonment of the mine last century.

A few veins, particularly in the Coniston and New-lands areas, contain small but significant quantities of other copper sulphides such as chalcocite, digenite, djurleite and bornite. Some of these, at least locally, may indicate secondary enrichment of a primary sulphide assemblage.

Until comparatively recently these deposits were tentatively regarded as the products of an early Devonian mineralising event. However, recent investigations have demonstrated that, as they were cleaved along with the wall-rocks, they must have formed during or shortly after the final phases of Ordovician volcanic activity. The Borrowdale Volcanic Group rocks are likely to have been the source of copper and other metals, with formation temperatures within the veins ranging from around 400 to 200°C.

Unlike copper-bearing veins elsewhere in Britain, particularly south-west England, the veins of the Lake District generally do not exhibit a wealth of colourful supergene minerals in their near surface levels. In part this may be due to the removal of oxidised assemblages during the Quaternary glaciations. Small amounts of malachite and chrysocolla occur in a few veins at Coniston and Newlands, but the greatest concentration of such assemblages within the Lake District are those of the Caldbeck Fells.

Lead-Zinc Veins

Veins which carry lead and zinc minerals as the major metallic constituents, like the copper-bearing veins, comprise an important group of the Lake District's mineral deposits. These too were formerly mined on a large scale. The Lake District includes Britain's third richest lead mine at Greenside, near Ullswater, which closed as recently as 1962 with a record of approximately 200,000 tons of lead concentrates to its credit. Zinc mining was never on such a scale, but only in 1990 did active exploration for zinc end with the final abandonment of Force Crag mine near Keswick. Other major concentrations of lead-zinc mineralisation include those at Thornthwaite, Threlkeld, Caldbeck Fells, Vale of Newlands, Brandlehow, Helvellyn, Hartsop and Eagle Crag.

In these veins, galena and sphalerite are the main ore minerals, accompanied locally by minor amounts of chalcopyrite. Silver is an important impurity in much of the Lake District galena, with up to 30 ounces of silver per ton[*] (840 ppm) of lead recorded from parts

CHALCOPYRITE CRYSTALS

From Coniston, Cumbria. Small twinned and single iridescent crystals scattered on rhyolitic matrix. This specimen (c.17.5cm across) [BM 1985, MI.12359] is from the H. Ludlam Collection. Ludlam (1822-80) was a London hosier and mineral collector who acquired the collections of Charles Turner and William Neville. His collection was bequeathed to the Museum of Practical Geology and passed to the Natural History Museum in 1985. (©NHMPL)

of the Caldbeck Fells. Most of the galena also contains an abundance of microscopic inclusions of native antimony and antimony sulphosalts such as bournonite and tetrahedrite. The latter mineral occurs locally in some abundance, although not in economic quantity, at a handful of localities such as Eagle Crag mine, Patterdale, where beautiful tetrahedral crystals up to about 5 mm across have been found.

Gangue minerals in these veins are mainly quartz, barite, calcite and dolomite. Fluorite is common at Brandlehow near Derwentwater and at Whitecombe Beck, Black Combe. It has also been found at Hartsop Hall and the deeper levels of Force Crag mine. Quartz pseudomorphs and epimorphs after barite are common at many localities, suggesting that barite was originally a major constituent of some of these veins.

---

* Silver recovery (or assay) figures recorded by the mining companies was usually quoted as ounces per 'ton', i.e. an imperial ton rather than the metric tonne employed today. This convention is retained in this book when referring to such silver values.

ABOVE: GREENSIDE MINE GALENA

Galena crystals from Greenside mine, Glenridding, Cumbria. This specimen, which is approximately 30 cm across, is from the mineral collections of the Department of Earth Sciences, University of Durham. (©NMS)

RIGHT: SPHALERITE CRYSTALS, FORCE CRAG MINE

Large black sphalerite crystals partly encrusted with honey-yellow siderite, collected in 1985 from a sublevel beneath the No. 0 Level, Force Crag mine, Braithwaite, Cumbria. This mine was a noted source of fine specimens of this sort. This specimen is 5 cm high. (©NERC)

A few Lake District lead-zinc veins show evidence of depth zonation. At Force Crag, abundant barite with some manganese oxides gives way at depth to sulphide-rich assemblages with common siderite and only sporadic concentrations of barite. The Greenside vein showed a similar downward decrease in barite content, accompanied by an increase in chalcopyrite in the lowest worked levels.

The Lake District lead-zinc veins are considered to have been deposited from highly saline fluids which may have derived metals from Skiddaw Group sediments, the Lake District Batholith, or from the Carboniferous sediments of the Northumberland-Solway 'trough' to the north of the Lake District. Emplacement of these veins during Late Carboniferous or early Permian times has been suggested. Fluid inclusion studies indicate formation temperatures in the range of 130-110°C for some veins. These, and other mineralogical and geological similarities between the lead-zinc veins of the Lake District and those of the nearby northern

Pennines, suggest that these deposits may share a common origin. Sulphur isotope studies, which suggest derivation of the sulphur in the barite in both the Lake District and northern Pennine veins, from Carboniferous evaporites, adds weight to this suggestion.

Like the Lake District copper veins, the lead-zinc veins are not noted for their supergene assemblages, except in the Caldbeck Fells where a wealth of species such as anglesite, cerussite, leadhillite, pyromorphite, mimetite, caledonite, linarite and many others are well known. These species have been described and illustrated in this book's companion volume on the Caldbeck Fells (Cooper and Stanley [1990]). The considerable supergene alteration in this area is probably not simply related to the present zone of weathering. It has been proposed that some of these mineral assemblages may date from supergene processes as early as the Jurassic. It has also been suggested that at least some of the finely crystallised pyromorphite group minerals may be the products of near-surface primary mineralisation.

Barite Veins

In addition to its occurrence as a gangue mineral in the Lake District's lead-zinc veins, barite occurs in great abundance, without significant metal ores, in a number of veins in the Eycott Volcanic Group and Lower Carboniferous rocks of the northern Lake District. Barite is commonly the main constituent, although quartz and, in places, manganese oxides are also present: sulphide minerals are generally rare or absent. Examples include some of the east-west veins worked for barite at Potts Gill mine, and the veins of the Ruthwaite and Gilcrux mines in the Carboniferous rocks north of Cockermouth. Their presence in Carboniferous rocks indicates a post-Carboniferous, or at the very earliest, Late Carboniferous age for these veins. The distribution of these veins in a relatively narrow belt close to the Maryport-Gilcrux fault, which marks the northern margin of the Lake District massif, has been cited as evidence in support of an important zone of mineralisation along this line. Structural and stratigraphical similarities between this area and parts of Central Ireland, where large barite and base metal deposits occur within the Lower Carboniferous Limestone, suggest that similar, as yet undiscovered, deposits may exist at depth in Cumbria.

Gold Occurrences

Within the Lower Palaeozoic outcrop of the central Lake District, gold has been reported from over 20 separate localities. A handful of these are 19th-century records for which no supporting specimens can be traced. There are published references to the presence of gold in the sulphide ores of Brandlehow, Derwentwater; God's Blessing, Coniston; Hartsop Hall, Patterdale; and Goldscope, Vale of Newlands mines. Indeed, Queen Elizabeth I is said to have claimed the latter as a royal mine on account of the gold and silver content of the ore. However, the name 'Goldscope' is thought to be a corruption of the German name 'Gottesgab' ('God's Gift'), rather than an allusion to any precious metal content. The nature of the gold and the amount present in any of these ores has never been established. Microscopic grains of native gold have been identified recently in detailed studies of sulphide ores from Dale Head mine, Vale of Newlands, and Seathwaite mine, Coniston.

The majority of reported occurrences are based on specimens in the Natural History Museum collected by A. W. G. Kingsbury in the 1950s. Some of the specimens, particularly those from Carrock Fell mine and Embleton Quarry, are remarkably rich. However, recent examination of all Lake District gold specimens in the national collection casts doubt on the authenticity of many of them. It is also noteworthy that the only Kingsbury Lake District gold locality which has been independently confirmed by later collecting is the occurrence at Carrock Fell mine and at the nearby streams.

Studies of panned concentrates of stream sediments in recent years have revealed gold to be widely present in very small amounts in the Cockermouth area. The gold is thought to have been derived, at least in part, from glacial drift, possibly of southern Scottish origin, although it is possible that some may have originated more locally, perhaps from minor sulphide mineralisation within Carboniferous rocks.

The Northern Pennine Orefield

The northern Pennine orefield may be seen as two separate but related fields: the Alston orefield of Cumbria, Northumberland and Durham in the north, and the Askrigg orefield of Yorkshire in the south. Although these fields have many structural and mineralogical features in common, there are also important differences. Both may be regarded as part of the Mississippi Valley – Alpine type. The essential features, which are common to both orefields, will be briefly reviewed here before the characteristics of each field are outlined in slightly more detail.

The Pennine orefields have been major producers of lead ores with appreciable quantities of silver recovered as a by-product during smelting. Significant amounts of zinc ores, both sphalerite and smithsonite, have been raised, particularly in the Alston area where a small tonnage of copper ore has also been worked. Iron ores were mined on a large scale, particularly in Weardale. Since the demise of large-scale lead mining towards the end of the 19th century, the Pennine deposits have been important producers of such gangue or 'spar' minerals as fluorite, barites and, in places, witherite. (It should be noted that the terms 'fluorspar' and 'barites' are normally applied in the industry to the commercial product from the mines composed mainly of the minerals fluorite and barite.) Much research has been carried out on the north Pennine orefield, particularly the Alston part: descriptions of the geology and mineralisation are listed in the Bibliography.

It is clear from the broad outline of Carboniferous geology already given that the Alston and Askrigg 'blocks' are structural units characterised by comparatively thin Carboniferous sequences overlying Lower Palaeozoic rocks which were intruded by Late Caledonian granites. The 'block' areas are bounded by major fault belts, many of which were active during Carboniferous sedimentation. The gently inclined Carboniferous rocks on the blocks are cut by conjugate systems of mainly normal faults, many of which are mineralised as veins.

A feature common to both of the Pennine orefields described here is the great influence that wall-rocks have on the form and mineral content of the deposits. Where faults cut hard rocks such as limestones and sandstones, and in the Alston Block the Whin Sill, clean open fractures conducive to the formation of wide vein oreshoots developed. In soft or weak rocks such as shales, vein fractures are typically closed and barren. Where alternations of hard and soft rocks occur, as in the Yoredale sequences of much of the Pennines, veins commonly show vertical alternation of wide 'ribbon' oreshoots and barren intervals.

In addition to the typical fissure veins, the Pennine orefields are noteworthy for the abundance of replacement orebodies, known from their characteristic three-dimensional form as 'flats'. The metasomatic processes involved in this replacement result in an overall volume reduction in the affected limestone. The effect of this is to produce numerous cavities or vugs, usually lined with beautifully crystallised examples of the constituent minerals. Flats are particularly common in the Alston orefield from which large tonnages of lead, zinc and iron ore have been mined.

A characteristic feature of both orefields is the marked zonal distribution of the constituent minerals, most notably the gangue minerals. The zonal pattern within each of the fields will be outlined more fully below, but as a general rule it may be observed that deposits within the central part of each field carry

abundant fluorite. Surrounding this is a zone in which barium minerals are the main gangue. Whereas this pattern of distribution appears to have been well known to miners from an early date, the significance of this in understanding the origin of the deposits was not appreciated until about 70 years ago. In a study of the Alston Pennines, the late Sir Kingsley Dunham compared this zonation of minerals in the Pennines with the zonation of minerals in the veins surrounding the granites of south-west England. By analogy with these deposits he advocated the presence of a granite at depth beneath the Pennines as a source of the mineralisation. Support for this hypothesis came in the 1950s when geophysical surveys revealed strong negative gravity anomalies beneath the Alston and Askrigg blocks, the form of which appeared to coincide closely with the observed mineral zoning. Studies of the volatile contents of coals within the Alston area added weight to this idea by suggesting that coals in the centre of the orefield, and closest to the centre of the northernmost postulated granite, had been heated to the highest temperature. Drilling of the Rookhope Borehole in Weardale in 1960-61 and the Raydale Borehole in the Askrigg area in 1973, proved the presence

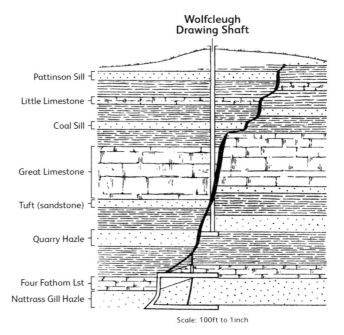

SECTION OF WOLFCLEUGH MINE, ROOKHOPE, CO. DURHAM

The steeper and wider portions of vein between sandstone and limestone wall-rocks, compared with the more gently inclined and narrow widths in shale, can be clearly seen. (Source: Carruthers and Strahan [1923] ©NERC)

of Late Caledonian granites beneath the Carboniferous rocks at each site. However, both granites were clearly pre-Carboniferous; the Weardale Granite in the Rookhope Borehole has a weathered top beneath the basal Carboniferous beds. Radiometric dating has given ages of 410 and 400 million years for the Weardale and Wensleydale Granites respectively. A second proving of the Weardale Granite was made at Eastgate, Weardale, in 2004 as part of an investigation of potential geothermal resources.

Clearly, neither granite could have been the primary source of the mineralisation. The generally accepted view of the origin of the Pennine mineralisation involves the transport of metals in solution in saline fluids which circulated convectively, driven by comparatively high heat flows from the concealed granites. The metals and other elements required to form the deposits were almost certainly derived by leaching from Lower Palaeozoic basement rocks, the granites, the Carboniferous sedimentary rocks and, in the Alston Pennines, the Whin Sill. Cooling of the mineralised fluids as they moved outwards from the emanative centres, together with reaction with other groundwater, accounts for the zonal arrangement of the constituent minerals.

Studies of fluid inclusions from vein minerals suggest that in the central parts of the Alston field fluorite crystallised from fluids at up to about 180°C: barite may have crystallised at temperatures as low as 50°C. Isotopic evidence for sulphur in barite, in both the Lake District and northern Pennines being derived from Carboniferous evaporites in the Northumberland-Solway trough, has already been mentioned. Much, if not all, the Pennine mineralisation appears to have been emplaced very soon after the intrusion of the Whin Sill in Late Carboniferous or Early Permian times.

The Alston Block

Of the two northern Pennine orefields, that of the Alston Block has been the most productive of mineral products and has attracted the most research interest on the genesis of the ore deposits.

Whereas it shares with Askrigg orefield to the south the same overall structural framework, there are important differences. Of particular relevance to the mineral deposits is the more varied nature of the Car-

ABOVE: SECTION OF 'FLATS' AT BOLTSBURN MINE

Section of Boltsburn vein and 'flats' at Boltsburn mine, Rookhope, Co. Durham. (Source: Carruthers and Strahan [1923] ©NERC)

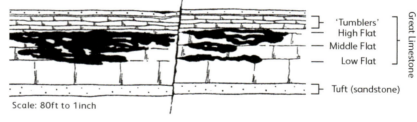

RIGHT: ROOKHOPE BOREHOLE

The drilling rig in 1960. (Photo courtesy of the Department of Earth Sciences, University of Durham)

FAR RIGHT: THIN SECTION OF WEARDALE GRANITE

From 427.5m depth, in the Rookhope Borehole. Large grey grains with patchy traces of cross-hatching are microcline; irregular black, grey and white grains are quartz; brightly coloured grains are muscovite. (C. H. Emeleus: University of Durham)

boniferous succession. Apart from the lowest part of the succession, limestones are much less abundant than farther south, with mudstones and sandstones comprising a much higher proportion of the sequence. The dolerite of the Whin Sill underlies much of the area, and crops out along the Pennine escarpment and in Teesdale.

The area is cut by a well-developed conjugate set of mainly normal faults, many of which have been mineralised. One of the field's most remarkable features is the concentric zoning of constituent minerals which makes it one of the world's finest examples of a zoned orefield. Veins within a central zone almost all contain abundant fluorite and quartz. Throughout much of the zone, purple of various shades is the dominant colour of the fluorite, although in a few places beautiful green crystals are found. Pale to deep yellow crystals are characteristic of the outer parts of the zone. Almost all of the fluorite within the Alston orefield exhibits strong to moderate fluorescence. Both colour and fluorescence appear to be the result of minute contents of rare-earth elements. Within the fluorite zone local concentrations of chalcopyrite, together with traces of bismuthinite and rare synchysite, monazite, xenotime

and at one locality cassiterite, may mark emanative centres of mineralisation. A few fluorite-bearing veins pass at depth into veins with abundant quartz and iron sulphides such as pyrrhotite. The Great Sulphur Vein of Alston Moor, which is distinguished by carrying large amounts of iron sulphides, including pyrrhotite, may represent the deep, 'root zone' of such a major fluorite-bearing vein, the upper levels of which have long been removed by erosion. Galena is common in the veins of the fluorite zone, but reaches its greatest abundance in deposits toward the outer part of the zone. Most of the galena is argentiferous with recorded silver values mainly between 4 and 8 oz per ton of lead metal. In a few places, galena with silver contents as high as 90 oz per ton of lead have been found. A description of silver minerals from Tynebottom mine, Garrigill, was based on specimens in the A. W. G. Kingsbury collection, although it now seems likely that the specimens were erroneously labelled from this locality. Sphalerite, usually a dark brown or black, iron-rich variety, is common, although the greatest concentrations of this mineral occur in a few areas near the margin of the fluorite and barium zones, particularly around Nenthead and in parts of Teesdale.

FLUORITE WITH FLUID INCLUSIONS

A clear cube of fluorite, approximately 2 cm across, with distinct purple and green colour banding and a large macroscopic fluid inclusion with a prominent large bubble. The specimen [NHMPL spec. no. BM 56270] is simply labelled 'Weardale'. (©NHMPL)

deposited between 200°C and 100°C and barite between 130°C and less than 50°C. The highest temperatures were reached at the centre of the mineralised area, although the highest individual temperatures occurred mainly at vein intersections. The heat source required for the hydrothermal solutions is thought to be related to radiogenic heat within the Weardale Granite after it was subsequently blanketed by sediments.

A NOTE ON FLUID INCLUSIONS

The study of fluid inclusions, 'cavities in minerals filled by trapped fluid', has enabled a great deal of information to be obtained on the nature of the original mineralising solutions in hydrothermal ore deposits. This is nowhere better shown than in the fluorite deposits of the northern Pennine orefield; especially large fluid inclusions occur in fluorite from Heights mine and other parts of the Weardale area. Studies on fluorite from Groverake mine and elsewhere have shown that fluid inclusions may be used as a possible prospecting tool for the ore mineralogist.

Generally, in the northern Pennine orefield, fluid inclusion studies point to moderately saline fluids with sodium-potassium ratios less than those of sea-water. Stable isotopes indicate that the fluids were mainly deep circulating formation waters which modified their composition by wall-rock reactions at depth. Fluorite was

Outside the fluorite zone the veins are dominated by a gangue of barium minerals. The transition from fluorite to barium zones in this orefield is very sharp. Only in a very small number of deposits, for example those of the Hilton and Murton mines on the Pennine escarpment, are fluorite and barium minerals found together in the same deposit. Within the barium zone, barite is generally the most abundant mineral. However, a unique feature of this orefield is the widespread occurrence of barium carbonate minerals, the most common of which is witherite, which was first recognised as a new species from specimens collected from Alston Moor. Several veins in which witherite is the main constituent have been worked commercially for the mineral. Notable examples include the Settlingstones vein, near Hexham, in the northernmost extremity of the orefield, and the South Moor vein in the Coal Measures of the adjoining Durham coalfield. In these mines run-of-mine ore commonly contained up to 90% barium carbonate. The double carbonates of barium and calcium, barytocalcite and alstonite, are also found, the former in some abundance, at several localities. Both of these minerals were first recognised as new species in the orefield in the 19th century. Although since found at a handful of other localities throughout the world, they both remain very rare minerals. The reason for their comparative abundance in the northern Pennines is not known. Where barium carbonate minerals have been subjected to the action of sulphate-bearing groundwater, either near the surface in supergene weathering or as a result of early alteration in the vein's development, barite with a very distinctive morphology has formed. Crystals of this barite are typically sharply pointed with the dominant faces (110) combined with (001). The crystals are often aggregated into 'coral-like' forms. In a few places where witherite veins cut the Whin Sill, small beautiful cruciform twinned crystals of the barium zeolite harmotome line vugs. Strontianite has been recorded from the witherite vein at Settlingstones.

Unusual concentrations of nickel minerals occur at a few places in the outer part of the orefield. A very small but rich deposit of nickeline with traces of ullmannite was extracted from the witherite vein at Settlingstones in the 1920s. Nickeline, accompanied by gersdorffite and millerite, has also been discovered at Hilton mine on the Pennine escarpment, and is prominent in a unique magnetite skarn assemblage at Lady's Rake mine in Upper Teesdale. This assemblage is important as it provides evidence that sulphide mineralisation began very soon after the emplacement of the Whin Sill and probably whilst it was still hot. Traces of cobalt and nickel sulpharsenides, tetrahedrite and arsenopyrite are common in the sulphide assemblage of the Alston orefield, a feature which serves to distinguish this orefield from that of Askrigg to the south.

Another characteristic feature of the Alston field is the great abundance of iron carbonate minerals. Siderite is common, together with ankerite of a wide range of compositions. Whereas these minerals are

REDBURN MINE, ROOKHOPE

Red Vein exposed in the No. 6 West Stope on the 17 Fathom Level. The vein consists mainly of massive purple and green fluorite. Prominent post-mineralisation slickensides follow narrow ribs of galena near the centre of the vein. Fine-grained brown sphalerite, which cements brecciated fluorite, may represent a later phase of mineralisation. Bands of white quartz locally cut the fluorite. Early quartz, with pyrite and chalcopyrite, occurs near the right-hand wall. (R. F. Symes)

LEFT: SCRAITHOLE VEIN,
SCRAITHOLE MINE, WEST
ALLENDALE

The vein, here in the Great Limestone, is composed mostly of
witherite with small amounts of
galena and numerous large blocks
of limestone wall-rock. Photo-
graphed in 1981. (T. F. Bridges)

ABOVE: ALSTONITE FROM
BROWNLEY HILL MINE, ALSTON,
CUMBRIA

White pyramidal crystals up to
5mm long, partly coated with
pyrite. (NMS G2004.44.4;
photo. © NMS)

common in veins throughout the orefield, they are
especially abundant in the replacement deposits in
limestone known locally as 'flats'. In these deposits
large volumes of limestone adjacent to veins have been
selectively replaced with iron carbonate together with
sulphides and other gangue minerals. Flats are most
common in some limestones, and within these, particular beds have been more susceptible to replacement.
The replaced rock within the flats is most usually a
compact crystalline aggregate of siderite and/or
ankerite, although fluorite, barite and silica, either as
quartz or chalcedony, may also be common and in
places dominant.

Many of the iron-rich flats, especially where the
carbonate minerals have been wholly or partly oxidised,
have provided large tonnages of iron ore. Bands and
pockets of sulphides, mainly galena and sphalerite, are
common in many flats where the sulphide content
often proved to be greater than in the adjoining veins.
Cavities, known locally as 'vugs' or 'loughs', are common and are typically lined with beautifully crystallised galena, sphalerite, quartz, fluorite, barite, calcite
or ankerite. Some of the orefield's finest specimens of
crystallised minerals have been collected from such
vugs. The abundance of iron carbonate and quartz in

the Alston orefield has been attributed to the scavenging
of iron and silica from the Whin Sill by the mineralising fluids.

Fluorite and barite are locally comparatively common in small quantities within the Magnesian Limestone of eastern County Durham. Of particular note
is an occurrence of colourless fluorite in the form of
highly unusual skeletal crystals in cavities within
dolomitic limestone in an abandoned dolomitic limestone quarry at Newton Aycliffe. Copper mineralisation, mainly in the form of chalcocite and azurite, with
traces of cuprite and native copper, has been found in
the Magnesian Limestone at Raisby Hill. All of these
occurrences may be regarded as outlying expressions
of the main northern Pennine mineralisation.

Supergene processes have produced the normal
oxidation products in the upper levels of veins and
flats. Large tonnages of goethite have been mined and
quarried from such deposits. Zinc and lead ores have
suffered extensive oxidation, particularly in the Middle
Fell area near Alston. Smithsonite is particularly common here where it was worked, as 'calamine', from
several mines. Although mostly in the dull brown form
known from its appearance as 'dry bone ore', beautiful
solid crystalline mammillated masses, of a pale green

THIN TABULAR EPIMORPHS FROM NENTSBERRY MINE

Thin tabular epimorphs, possibly after marcasite, in pyrite and siderite, from Nentsberry mine, Nenthead, Cumbria. This specimen, which is about 25 cm across, was collected by the late Sir Kingsley Dunham and described in his PhD thesis in 1932. (©NMS)

or yellow colour, were obtained from the Farnberry and Holyfield mines. The authors have seen polished specimens, said to be from here, used as an ornamental stone. Cerussite was mined as the main lead ore at Hudgill Burn mine. Spectacular specimens of this mineral have been found in Weardale at Stanhopeburn, Redburn and Stotfield Burn mines and elsewhere.

Other supergene minerals of note include erythrite, devilline and wroewolfeite at Tynebottom mine, Garrigill; anglesite, rosasite, aurichalcite, leadhillite and pyromorphite at Closehouse mine, Middleton-in-Teesdale; coronadite at Sedling mine, Cowshill; epsomite, serpierite and namuwite at Smallcleugh mine, Nenthead; anglesite at Pike Law mines, Teesdale; azurite and malachite from Cornriggs mine, Garrigill; brianyoungite from Brownley Hill mine, Nenthead; and hydrozincite and cadmium sulphide minerals from a number of mines in Alston Moor and Teesdale.

The Askrigg Block

Although this area has many similarities to the Alston Block, there are significant differences. Like the Alston field, this is an area of comparatively thin Carboni-

ferous rocks underlain by Lower Palaeozoic rocks in which occurs the Late Caledonian Wensleydale Granite. Limestone comprises a significant proportion of the Lower Carboniferous sequence, and cherts are abundant in the Namurian rocks, together with thick grits which are particularly prominent in the south, where they rest unconformably on the limestones. No post-Carboniferous intrusions equivalent to the Whin Sill occur in the Askrigg field. As in the Alston orefield, veins occupy a conjugate system of normal faults, although productive veins are concentrated in two main areas: the Swaledale and Wensleydale area in the north, and around Grassington and Greenhow in the south. Smaller, scattered deposits occur elsewhere.

The veins of the Askrigg orefield, like those of the Alston field, carry galena as their main ore mineral in a gangue of fluorite and barium minerals. Except in a few localities, sphalerite is much less abundant than in the Alston field and iron carbonates and silica minerals are generally rare. Flat deposits replacing limestone are present in the Askrigg field, but are less common than farther north.

Whereas a zonal distribution of gangue minerals can be distinguished, it is not as clearly developed as in the Alston field. Unlike the Alston orefield, where fluorite occurs in one main central zone almost invariably unaccompanied by barium minerals, in the Askrigg field it is found in a number of separate zones where almost everywhere it is inter-banded with witherite or barite. Fluorite in the Askrigg field is typically colourless, white or pale yellow: strong purple and green colours are very uncommon. Inclusions of bravoite, pyrite and marcasite are locally common in the fluorite.

Barite and witherite are very common, both within the fluorite zone and in a wide surrounding zone which corresponds with the barium zone of the Alston field. Barytocalcite is present in small amounts locally, although it has not been found in the same abundance as further north. Barite with the highly distinctive secondary morphology described above in the Alston area is extremely common in parts of the Askrigg field, suggesting that barium carbonate minerals may formerly have been more abundant than at present, or that they may be major constituents of the deposits at depth beneath the zone of supergene alteration. Several veins of the Askrigg field, especially in the Swaledale and Pateley

Bridge areas, carry a little strontianite. Surrounding the barium zone of the Askrigg orefield is a wide outer zone in which calcite is the principal gangue mineral.

Unlike the Alston orefield, quartz is rare or absent from all but a handful of deposits in the Askrigg field. Iron carbonates too are rare, although true dolomite, a mineral rarely found in the Alston orefield, is locally common.

Galena is the main ore mineral, with silver values generally less than 4 oz per ton of metallic lead. As in the Alston field, galena values tend to be highest around the outer margin of the fluorite zone. Whereas sphalerite is common today only in a few places, it may formerly have been present in some abundance as the zinc supergene minerals hemimorphite and smithsonite are common at some localities. A remarkable group of deposits of smithsonite were worked at Malham where the mineral, known locally as 'calamine', occurred filling caves in limestone.

It is perhaps worth explaining here the meaning of the term 'calamine' and the confusion which has arisen over its use. 'Calamine' is the ancient name for the 'dry bone' and similar varieties of smithsonite which were worked for brass manufacture. In their standard work on British mineralogy, Greg and Lettsom (1858) followed earlier authors in applying the term 'calamine' to zinc carbonate, and smithsonite to zinc hydroxysilicate, but pointed out that some European mineralogists had applied the name 'smithsonite' to the carbonate. They added that as Smithson, after whom the mineral was named, carried out much work on silicates, it was more appropriate that his name should be attached to zinc hydroxysilicate. However, international usage has not followed this view, and 'smithsonite' is today the name applied to the carbonate: 'hemimorphite' is the hydroxysilicate. Care is thus needed when reading old references to 'calamine' which may refer to either mineral.

As in the Alston field, near surface oxidation processes have produced small amounts of supergene minerals, notable amongst which are smithsonite and hemimorphite at numerous localities; the rare zinc silicate mineral fraipontite from Virgin Moss mine, Wensleydale and at Copperthwaite vein in Swaledale; aurichalcite, rosasite and jarosite group minerals from Grassington Moor; cinnabar at several localities in Swaledale and around Greenhow; and cinnabar, hemimorphite, prosopite, doyleite and otavite from Coldstones Quarry, Greenhow Hill. Spectacular examples of the 'flos ferri' variety of aragonite, of probable supergene origin, have been found in caverns at Sloat Hole, Faggergill and Windegg mines in Arkengarthdale. Specimens of the rare zinc phosphates parahopeite and spencerite, native mercury and calomel, said to have been obtained from Grassington Moor by A. W. G. Kingsbury, are almost certainly incorrectly labelled by their collector.

A remarkable cluster of copper deposits was worked for a short period, mainly in the 18th century, in and around the village of Middleton Tyas, south-west of Darlington. The deposits consisted of veins, flats and 'pipes' developed predominantly in the Undersett Limestone, which crops out here within the Middleton Tyas anticline. The precise form of the deposits is not known as the workings have long been abandoned and inaccessible. Chalcopyrite and djurleite appear to have been the primary ore minerals. Digenite, bornite and covellite, which were locally common, suggest secondary enrichment. Oxidation minerals include abundant malachite and azurite with, in places, cuprite and native copper. Contemporary accounts suggest that grades were extremely rich, with recoveries of up to 66% copper recorded from some ores. A similar assemblage of copper minerals was found a few years ago within the Main Limestone at Forcett Quarry about 9 km WNW of Middleton Tyas.

Studies carried out on the relatively meagre amounts of surviving mineralised material suggest these small concentrations of remarkably rich mineralisation result from supergene enrichment of copper sulphide mineralisation within the outer part of the Askrigg orefield. Other small veins with primary copper mineralisation occur a few kilometres further south near Richmond.

A few clusters of copper-bearing veins within Lower Carboniferous rocks occur close to the Pennine escarpment in the Kirkby Stephen area. Like the veins of the Middleton Tyas and Richmond areas, these appear to form part of an outlying group of veins within the Askrigg orefield. Although trials have been made on most of these, none has produced more than a very small amount of ore. Chalcopyrite is locally associated with galena in a gangue of quartz and, in places, fluorite and barite. At Stennerskeugh Clouds massive tennantite is common; at Cumpston Hill tetrahedrite is a major sulphide. Several of these localities have yielded small specimens of beautifully crystallised azurite, accompanied by small examples of cupro-adamite at High Longrigg, near Kirkby Stephen.

A number of small copper veins with conspicuous azurite and malachite were also worked in the Lower Carboniferous limestones of the Malham area.

The West and South Cumbrian Iron Orefields

The Lower Carboniferous limestones of west Cumbria and the Millom and Furness areas of south Cumbria contain numerous, often very large, bodies of hematite. These are typically vein-like bodies or large irregular, commonly flat-lying, deposits which, like the flats of the northern Pennines, comprise extensive replacements of limestone, usually adjacent to faults. The orebodies are composed mainly of hard, compact, massive hematite. As in the flats of the northern Pennine orefields, replacement of limestone by hematite has resulted in a volume reduction with many open voids or vugs, usually known in west Cumbria as loughs, within the orebodies. Typically these are lined with small lustrous black crystals of specular hematite, but also common within them are mammillated masses of the hematite variety known as 'kidney ore' for which the Cumbrian mines have long been famous. 'Kidney ore', which also occurs as irregular bands within the massive ore, consists of radiating fibrous crystals of hematite elongated perpendicular to the c-axis. When broken the 'kidneys' tend to break into conical fragments known to the miners as 'pencil ore'.

In addition to hematite, manganese oxide minerals have been found in a number of orebodies, though always in very small amounts. A small tonnage of a manganese ore which included such minerals as haus-mannite, manganite, pyrolusite and rhodochrosite, found near the margin of the hematite orebody, was worked at Wyndham mine, Bigrigg. Other metallic minerals are generally rare, though small amounts of

galena and chalcopyrite have been identified locally. Some copper ore, consisting mainly of chalcocite, was worked alongside hematite from the Anty Cross mine at Dalton-in-Furness, though it is likely that the copper vein pre-dated the hematite mineralisation.

Although non-metalliferous gangue minerals are generally present in only very subordinate amounts, the Cumbrian iron ore mines have long been celebra-ted for beautiful specimens of quartz, calcite, aragonite, barite, fluorite and dolomite recovered from vugs (loughs). Barite is locally common as a contaminant of hematite near the margins of some orebodies, and dolomite is a common replacement of limestone adjacent to many orebodies.

Intensely dolomitised Lower Carboniferous lime-stones adjacent to faults and joints, at several places around the margins of the Lake District, have long been known for beautiful groups of aragonite crystals, accompanied locally by good calcite, a little hematite and traces of manganese oxides. This mineralisation, which has been exposed at a number of quarries including those at Eskett and Kelton in west Cumbria, and Parkhouse north of Penrith, may be related to the hematite mineralisation.

Fissure veins of hematite, also with only very minor quantities of gangue minerals, occur within the Lower Palaeozoic rocks of the central Lake District. The largest and formerly most important economically, were those within Skiddaw Group mudstones at the Knockmurton mines north of Ennerdale Bridge. Other

KIDNEY ORE, FLORENCE MINE

Kidney ore exposed underground in the workings of the Lonely Hearts Orebody, Florence mine, Egremont, Cumbria, January 2007. (B. Young)

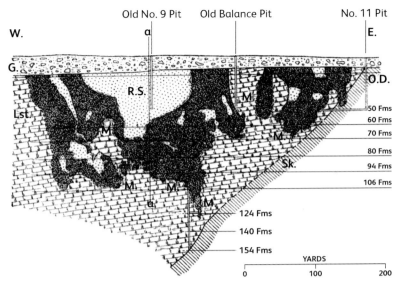

Old No. 9 Pit   Old Balance Pit   No. 11 Pit

W.   a   E.

50 Fms
60 Fms
70 Fms
80 Fms
94 Fms
106 Fms
124 Fms
140 Fms
154 Fms

YARDS
0   100   200

G.   Glacial beds
Lst.   Limestone
R.S.   Red sand, *etc.*
M.   'Muck' and 'hunger'
Sk.   Skiddaw Slate
     Hematite

SECTION THROUGH THE HEMATITE 'SOP'
ORE-BODY, BURLINGTON PIT, PARK MINES,
CUMBRIA

[Source: Smith (1924) ©NERC]

veins of this type occur in Ennerdale, Eskdale, at Red Tarn and near Grasmere. Their textural and mineralogical similarities clearly link them with those of the main hematite orefields within the Carboniferous limestone.

The origin of this mineralisation has long attracted attention and controversy. The most generally accepted views of the origin of these deposits suggest that the iron was derived from iron-rich Permo-Triassic sediments in the Irish Sea basin or from granites beneath the western Lake District. Convective leaching of this iron was effected by saline fluids circulating through permeable Permo-Triassic rocks in response to a heat source beneath the Irish Sea, or perhaps within the Lake District batholith. The iron-rich fluids were driven up-dip, gaining access via faults and fissures to the Carboniferous limestones where large-scale replacement of the limestone by hematite took place. The association of orebodies with the unconformable cover of Permo-Triassic rocks has long been known. Orebodies are typically found where permeable Permo-Triassic rocks, such as the St Bees Sandstone, or the breccias known locally as 'brockram', directly overlie the limestone. Where mudstones of Upper Carboniferous or Permian age intervene, orebodies are generally absent. This relationship suggests that mineralising fluids may have flowed, at least in part, through the permeable Permo-Triassic cover rocks: impermeable mudstones appear to have acted as a barrier preventing

access of these fluids to the limestones. The hematite veins in the Lake District formed where these fluids penetrated fractures within the Lower Palaeozoic rocks.

Similar large replacement orebodies occur in the south Cumbrian field, though this area also contains numerous hematite bodies known as 'sops', which are unique to the Furness area. Sops are typically circular to oval in plan and take the form of downward-tapering, roughly conical bodies composed mainly of brecciated hematite with a central filling of sand or sandstone rubble. The generally accepted explanation of sop formation suggests that they result from the deposition of hematite in large karstic dissolution cavities in limestone from the same fluids that elsewhere deposited hematite in replacement and vein-like bodies. Mineralisation took place beneath a cover of St Bees Sandstone, the collapse of which into the remaining space at the top of the developing sop produced the central core of sand and sandstone rubble. The characteristic brecciation of the hematite suggests that limestone dissolution may have continued during or after mineralisation.

Fluid inclusions within gangue minerals suggest that the mineralising fluids were concentrated saline brines at temperatures of up to 120°C. Arguments continue over the age of mineralisation, though early Permian, early Triassic or post-Triassic dates have all been proposed.

The Mines
and Minerals

FOR centuries, mining for a great variety of minerals has been one of the major industries of northern England.

To many, the north of England is inextricably linked with coal mining. The area covered by this book includes two of Britain's longest worked coalfields: the west Cumbrian coalfield centred around Workington, and the much more extensive Northumberland and Durham coalfield to the east of the Pennines with the cities of Newcastle-upon-Tyne, Sunderland and Durham near its heart. Over several centuries both coalfields have figured prominently in the development of many aspects of coal mining technology, including the development of the original flame safety lamps, the Davy and Stephenson lamps. Millions of tons of coal from both coalfields have, over the years, sustained such heavy industries as iron and steel making and ship building. From an early date a sea-borne coal trade flourished from east coast ports on the rivers Tyne and Wear. 'Coals from Newcastle' supplied the fires of London in the 17th and 18th centuries, and during the hey-day of steam navigation fuelled a major part of Britain's shipping. Recent years have seen the almost complete collapse of coal mining. The region's last major underground coal mine closed in 2005 and coal extraction is today limited to a handful of large opencast sites.

Whereas the region's coal mines have yielded some notable mineral specimens, the main mineralogical interest of northern England lies in the many mainly metalliferous mines.

Within northern England mineralisation is concentrated in a number of more or less discrete orefields. For convenience of description the region has been divided into a number of areas of interest. In the following pages brief comments are offered on a selection of the most significant, or best known, mineralised sites. It must be stressed, however, that these are by no means exhaustive accounts of each area or of the sites of interest within them. For more details of the geology, mineralisation and mining history, the reader should consult the appropriate texts listed in the Selected Bibliography.

The present pattern of administrative county boundaries dates back to the last major reorganisation of local government boundaries in 1974. Information accompanying specimens collected or labelled prior to this reorganisation may therefore not accord with the modern counties. For much of the Pennine area, boundary changes may affect the labelling of specimens from parts of southern County Durham and Yorkshire, as well as the former counties of Cumberland and Westmorland (now Cumbria). West of the Pennines, Cumbria incorporates all of the former counties of Cumberland and Westmorland, together with the former detached portion of Lancashire in the extreme south of the Lake District and Furness. Reference to a pre-1974 map of English counties is therefore recommended when studying older labels or records.

The Lake District

The unique scenery of the Lake District, England's largest National Park, has long provided inspiration for writers, poets and artists. This landscape of peaceful lakes and valleys set amidst spectacular and rugged mountains, today attracts visitors by the million. Few perhaps realise that in times past the Lakeland fells were the home of important metal mining and processing industries: metalliferous mining is arguably one of the Lake District's oldest traditional industries. Ores of antimony, arsenic, cobalt, copper, iron, lead, manganese, nickel, silver, tungsten and zinc have all been worked, some in large amounts, together with valuable quantities of non-metallic minerals such as barite and graphite. The legacies of this activity are readily apparent in a few places, for instance at Coniston, the Newlands Valley and Glenridding where the remains of old mines are essential, indeed often protected, elements in the landscape. Elsewhere, however, evidence of the dozens of long-abandoned mines and trial workings is much less conspicuous and must be sought out by the interested visitor. Interpretive displays of the district's mining history can be seen at Threlkeld Quarry, the Keswick Mining Museum and at the Ruskin Museum, Coniston.

Included within the Lake District is the small, but extremely important, mining field of the Caldbeck Fells.

As this area is justly famous, and has been the subject of a companion volume to this book (Cooper and Stanley [1990]), it will not be discussed further here.

Although the Lake District has a long history of mining, the early origins of the industry are obscure. Scattered bloomery sites give evidence of primitive iron smelting, perhaps as early as pre-Roman times. There is no definite evidence of mining by the Romans, although as they occupied the area, and built roads and camps here, it seems almost inconceivable that they did not encounter, and at least investigate, mineralised outcrops. Some scant documentary evidence exists for 13th, 14th and 15th-century mining, from which it seems that mining was well-established in parts of the Caldbeck Fells and Newlands Valley. References in the 12th century to the 'Carlisle Silver mines' may refer, in part, to some of the Lake District lead mines. Many, if not most, of the lead ores of the Lake District are known to be silver-bearing. Although assay values and recovery figures are not as comprehensive as those for the north Pennines, silver contents of between 800 and almost 1000 ppm (approximately 28 to 35 ounces per ton) are common. Silver is known to have been recovered from the lead ore raised at numerous Lake District mines. It is said that silver, extracted from lead mined at Goldscope mine in the Vale of Newlands, was used to make the Goldscope Loving Cup (see page 38), now held in Keswick Museum.

Reliable records of Lake District mining begin in the 16th century with the advent of exploration and mining by German miners. At this time Germany led the world in mining technology. German mining interests received considerable encouragement from the English Crown, anxious to promote domestic sources of metals in the interests of national security and as a means of increasing royalty payments to the Crown. At this time, under the privilege of Mines Royal, the production of gold or silver from a mine was a royal prerogative. It was in 1563 that Daniel Hochstetter, a mining engineer and smelting expert representing Haug Langauer and Company of Augsburg, arrived in England. With Thomas Thurland (Master of the Savoy Hospital), William Humphreys (Assay Master of the Mint), Thomas Smythe (Collector of Customs in the Port of London) and others, he was granted the right to prospect for metal ores in Cumberland, Cornwall,

The Lake District

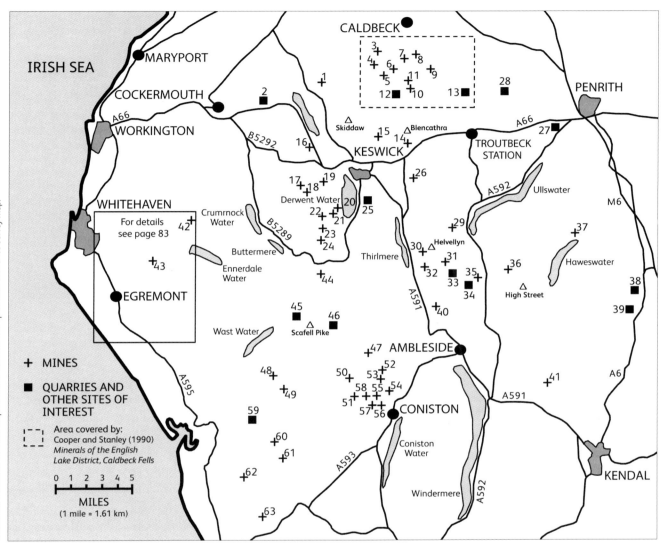

For reasons of clarity, only a selection of roads, rivers and settlements are shown. For more detailed location of individual sites this map should be used with an up-to-date Ordnance Survey map.

IRISH SEA

MARYPORT
COCKERMOUTH
WORKINGTON
A66
WHITEHAVEN
Crumrnock Water
Buttermere
Ennerdale Water
EGREMONT

For details see page 83

+ MINES
■ QUARRIES AND OTHER SITES OF INTEREST
┈ Area covered by: Cooper and Stanley (1990) Minerals of the English Lake District, Caldbeck Fells

0 1 2 3 4 5
MILES
(1 mile = 1.61 km)

CALDBECK
PENRITH
Skiddaw
Blencathra
KESWICK
TROUTBECK STATION
A66
Ullswater
M6
Derwent Water
Helvellyn
Thirlmere
Haweswater
High Street
Wast Water
Scafell Pike
AMBLESIDE
CONISTON
Coniston Water
Windermere
KENDAL
A595
A591
A592
A593
A6
B5292
B5289

1 Robin Hood mine	17 Force Crag mine	33 St Sunday Crag	49 South Cumberland mine
2 Embleton Quarry	18 Scar Crag (Cobalt) mine	34 Hogget Gill	50 Cockley Beck mine
3 Brae Fell mine	19 Barrow mine	35 Hartsop Hall mine	51 Seathwaite Tarn mine
4 Red Gill mine	20 Brandlehow mines	36 Low Hartsop mine	52 Greenburn mine
5 Roughton Gill mine	21 Yewthwaite mine	37 Burn Banks mine	53 Birk Fell Hause mine
6 Mexico mine	22 Goldscope mine	38 Shap 'Blue' Quarry	54 Tilberthwaite mine
7 Potts Gill mine	23 Castle Nook mine	39 Shap 'Pink' Quarry	55 Red Dell Head mine
8 Sandbeds mine	24 Dale Head mines	40 Grasmere mine	56 Bonsor mine
9 Carrock End mine	25 Walla Crag	41 Millrigg mine	57 Paddy End mine
10 Carrock Fell mine	26 Wanthwaite mine	42 Knockmurton & Kelton mines	58 Black Scar mine
11 Brandy Gill mine	27 Baron Cross Quarry	43 Kinniside mine	59 Water Crag
12 Wet Swine Gill	28 Parkhouse Quarry	44 Seathwaite (Graphite) mine	60 Hesk Fell mine
13 Linewath Bridge	29 Greenside mine	45 Lingmell Gill	61 Ulpha mines
14 Threlkeld mine	30 Helvellyn mine	46 Ore Gap	62 Buckbarrow Beck mine
15 Glenderaterra mine	31 Eagle Crag mine	47 Red Tarn mine	63 Whitecombe Beck mines
16 Thornthwaite mine	32 Birkside Gill mine	48 Nab Gill mine	

37

ABOVE: GOLDSCOPE MINE,
VALE OF NEWLANDS, CUMBRIA

Above the highest part of the large dumps at
this famous old Lake District mine, the course
of the vein may be traced up the hillside.
(B. Young)

RIGHT: THE GOLDSCOPE
LOVING CUP

This cup, now in Keswick Museum [No. 1483],
is said to have been made from silver extracted
from lead ores from Goldscope mine in the
Newlands Valley. (©NHMPL)

Wales and elsewhere. Initial results in Cumberland were encouraging, and in 1564 an agreement was concluded between Elizabeth I, Hochstetter and a number of English and German shareholders. Under this agreement the prospectors acquired the right to mine gold, silver, copper and quicksilver, with the Crown retaining the right to buy metals at advantageous prices. In 1568 Hochstetter and his associates were incorporated by charter as 'Governors, Assistants and Commonality of the Company of Mines Royal', usually known as the 'Company of Mines Royal'. These years saw the start of work on numerous Lake District mines, as well as the establishment of large copper smelting works at Keswick, fuelled with local peat, charcoal and coal from nearby. By the late 1570s the Company of Mines Royal was beset with a variety of financial difficulties. When Daniel Hochstetter died in 1591 his two sons took over management of the company, and by the beginning of the Civil War in 1642 the Keswick mines and copper works were closed.

A new impetus was given to Lake District mining in the 17th century when leases were taken by John Bathurst and David Davies to work mines in the Newlands Valley and the Derwent fells. Early in this century the famous Banks family acquired the Borrowdale graphite mine and eventually the Keswick pencil

industry was established. Mining activity continued during the 18th century, although by now lead had overtaken copper as the main metal sought.

The 19th century saw a further increase in mining throughout the district, aided by great improvements in the road system. Major developments at this time included the expansion of mining at Coniston and Greenside, with other important developments at Threlkeld, Thornthwaite and Force Crag. However, towards the end of the century a worldwide collapse in metal prices signalled an end for all but a handful of Lake District mines. Force Crag, Thornthwaite and Threlkeld mines all continued in production into the early years of the 20th century, but could not survive a further slump in metal prices at the end of the First World War. Only Greenside mine continued to work on any scale into the 20th century, finally closing in 1962. Attempts at reviving mining at Coniston were abortive. Several unsuccessful attempts were made in the 20th century to mine tungsten at Carrock Fell, the most recent ending in 1981. Mining of barite and zinc ore at Force Crag was important during the first half of the century, although attempts in the 1980s to resume commercial mining failed.

One of the greatest concentrations of old mines lies south-west of Keswick in the Newlands Valley. Here are some of the district's best known and oldest mines with rather sketchy documentary records dating back to the 13th century. At Goldscope mine, although there is evidence of earlier working, the first systematic exploration and working, mainly for copper, was begun by German miners under Hochstetter's direction in 1564. It was these miners who gave the mine its rather intriguing name. Originally thought to have been 'Gottesgab' (God's gift), this became corrupted locally to Gowd Scalp, Gold Scalp and eventually Goldscope. The name does not in any way derive from the very small gold content of some of the ore. It was this supposed precious metal content which led to a famous 16th-century dispute between the Earl of Northumberland, then owner of the mine, and the Crown who claimed it as a Mine Royal. The unfortunate Earl lost his case, his mine, and eventually his head, after his support for Mary Queen of Scots. Mining at Goldscope continued intermittently over succeeding centuries, during which time lead veins were discovered and worked. The most recent attempt, in 1917, to reopen the mine failed. The large silver cup, known locally as the Goldscope Loving Cup, said to have been made from silver extracted from Goldscope ores, has been mentioned above.

Further up the Newlands Valley, beyond the site of the small Castle Nook mine, is the cluster of old workings known collectively as Dale Head mine. Several narrow veins, including that known as the Long Work, were worked here for copper, some by the Elizabethan German miners. Open stopes and trenches, with some remaining vein exposures containing chalcopyrite and pyrrhotite in Skiddaw Group wall-rocks, are present in the valley floor. Higher up the fellside, below Dale Head Crag, occur other copper veins in Borrowdale Volcanic Group wall-rocks. Near the remains of a dressing floor, possibly of Elizabethan date, may be seen fragments of oxidised copper ores containing bornite, djurleite with crusts of malachite and chrysocolla.

On the east side of the Newlands Valley, at Little Town, the extensive dumps of Yewthwaite mine are conspicuous. This mine, which seems not to have been developed until the later years of the 18th century, worked both galena and cerussite from a roughly N-S vein. The same vein was worked for lead, from about

1680 to 1890, on the west side of the valley at Barrow mine, the scree-like spoil heaps of which are conspicuous on the hillside above Uzzicar.

Several lead mines were worked on the west side of Derwentwater on Catbells. The largest of these was the Brandlehow (sometimes referred to as Brandley) mine. Old opencast workings almost certainly pre-date the introduction of gunpowder in the early 17th century, but little or nothing is known of the mine's history until the 19th century when the so-called Salt Level was driven. This takes its name from a strong saline spring cut in the workings. These waters, and those from a similar spring at the nearby Salt Well mines at Manesty, where a small spa was briefly established, were occasionally used medicinally. Brandlehow mine closed in 1891. A short distance to the north, and almost certainly on the same vein, is Old Brandley mine. A feature of interest here is the abundance of fluorite which occurs here as colourless or pale yellow cubes in the upper workings. Also notable are well-formed quartz pseudomorphs after 'cock's-comb' barite and fluorite. Good specimens of bright green pyromorphite have also been found.

Force Crag mine at the head of Coledale, west of Braithwaite, is the Lake District's last worked metalliferous mine. Although lead ore was known from the head of Coledale in the 16th century, nothing seems to be known about mining here until about 1830 when mining for lead was begun by a Keswick-based company. Because of the abundance of barite in the vein, the mine was being worked for this mineral by 1860. In the early years of the 20th century, an Elmore flotation plant capable of separating galena, sphalerite and barite was installed, and production of these minerals continued under a succession of owners until the catastrophic post-Great War fall in metal prices forced its closure. The mine was reopened in 1930 and barites production continued intermittently under a variety of owners until 1966. Since then a number of attempts have been made to restart the mine, the most recent ending in the 1990s. Force Crag mine has long enjoyed a reputation for the quality of large lustrous black or dark brown crystals of sphalerite, typically accompanied by yellowish brown siderite. Fine specimens of these, and of beautiful white tabular barite crystals, were still being recovered when the mine finally closed.

OLD BRANDLEY MINE, DERWENTWATER, CUMBRIA

Old surface workings on the vein complex in Skiddaw Group wall-rocks. (B. Young)

FORCE CRAG MINE, BRAITHWAITE, CUMBRIA

The building housing the former treatment plant. (B. Young)

Small crystals of pale yellow fluorite were occasionally found near the walls of the vein. In the upper levels of the mine, fine black botryoidal psilomelane was abundant; a small amount is understood to have been sold as manganese ore. Although Force Crag was cited by Greg and Lettsom (1858) as the original British locality for stolzite, the mineralogy and geological setting of the vein make this extremely unlikely.

South of Force Crag, near the summit of Scar Crag, lie the very small workings of the notorious Cobalt mine. Considerable sums were raised to open and equip this mine in 1848, although little if any cobalt was ever recovered, and the project may have been little more than an opportunistic money-raising venture. Cobalt has been shown to occur here in very small amounts in alloclasite and skutterudite, which occur as small inclusions within abundant arsenopyrite. The vein is also of interest as a source of apatite crystals and small specimens of wavellite and variscite.

Thornthwaite mine, near the southern end of Bassenthwaite, was one of the district's largest lead and zinc mines, and one of the latest to survive until its closure in 1920 was forced by a collapse in world metal prices. The mine was once well-known to local collectors as a source of well-formed pyrite cubes.

Robin Hood Farm, north of Bassenthwaite, is the site of one of the Lake District's unique small mines.

Here in the 1840s a few tonnes of stibnite were mined from a shallow shaft. Little is known of the nature of the deposit, although it appears to have been a quartz vein. Good rich specimens from here may be seen in many collections. Apart from a drainage adit, in which there are no signs of mineralisation, little remains today to mark the site. Although this is the only Lake District locality at which antimony ores were worked, antimony-bearing veins are known from several other localities in the district. A complex assemblage of antimony sulphides occurs at Wet Swine Gill, near the Carrock Fell tungsten deposit. Antimony-bearing veins are known at Hogget Gill, Patterdale; and there are old but unsubstantiated records of stibnite-bearing veins on the nearby St Sunday Crags. Large glacial erratics of almost pure stibnite were found many years ago during the building of the Penrith-Keswick railway line at Troutbeck Station, although the *in situ* source of this rich ore remains unknown.

Threlkeld mine, at the southern foot of Blencathra, is another of the district's larger lead and zinc mines. Dating from pre-gunpowder days, the mine was worked on a considerable scale between 1879 and 1928, surviving the catastrophic collapse of world metal prices which forced the closure of Thornthwaite and Force Crag mines. Despite its size, the mine does not seem to have yielded many mineralogical specimens of interest, although a few fine examples of 'cock's-comb' barite were collected from the dumps some years ago.

Remotely situated at the head of Bannerdale is the small Bannerdale mine which was known to have been producing lead ore in the latter half of the 19th century. South of the lead mine, a tiny adit in Bannerdale Crags was driven on a bed of graphitic mudstone in the Skiddaw Group. This is said to have been worked as low grade graphite for making grate polish and pencil-making, but the amount obtained must have been very small.

Lead-bearing veins hosted by rocks of the Borrowdale Volcanic Group occur on the eastern side of the Helvellyn range of mountains. Foremost amongst these was Greenside mine, at Glenridding. Not only was this the Lake District's largest, most productive and longest worked lead mine, but it was one of Britain's most important lead producers with a total production of at least 200,000 tons of lead concentrates to its credit. The deposit was probably first discovered in the mid-

17th century, though little is known of its career until towards the end of the 18th century. From then until its abandonment in 1962 the mine worked almost continuously. In 1891 Greenside was the first British mine to install electrical winding gear underground. Shortly afterwards it became the first to employ an electric locomotive for underground haulage. Such was the mine's success that the company built rows of cottages for its employees in Glenridding and in Glencoyndale. A smelter, built at the mine in 1830, continued to process the ore until early in the 20th century. The N-S trending Greenside vein, which carried galena in a matrix of broken volcanic country rock and quartz, was up to 15 m wide in places, although widths of around 2.5 m were more normal. As the vein was followed downward into mudstones of the Skiddaw Group it turned barren, heralding the end of the mine. During its final years, abandoned stopes were used for seismic research into the detection of underground nuclear tests. A few fine specimens of crystallised galena, including some with a striking iridescent tarnish, were found here. Some well-formed 'cock's-comb' barite occurred, especially in the higher levels of the mine, and good specimens of crystallised calcite were occasionally found. During the 1980s some attractive specimens of blue mammillated hemimorphite, rather reminiscent of that from Roughton Gill mine in the Caldbeck Fells, were collected from an old adit on the vein high in Glencoyndale.

About two miles south of Greenside, Eagle Crag at the head of Grisedale is scarred by a narrow gully marking the course of the Eagle Crag vein. There is evidence here of Elizabethan mining, although most of the extensive surface dumps date from 18th and 19th-century working: the mine appears to have been last worked in about 1880. Eagle Crag vein is unusual amongst Lake District veins in that it contains abundant coarse-grained tetrahedrite, some in the form of fine bright euhedral crystals.

GREENSIDE MINE,
GLENRIDDING, CUMBRIA

Miners using a compressed air rock-drill in the late 19th or early 20th century. Drilling with these early machines was dry and produced huge quantities of dust. (Beamish)

Near the foot of Kirkstone Pass, Hartsop Hall mine worked a NNE-SSW trending lead vein from several levels. First worked in the 17th century, the most extensive activity was during the 19th century. Unsuccessful attempts were made to resume mining between 1931 and 1942. A feature of this vein is the presence of small amounts of yellow fluorite, a rare mineral in Lake District veins. Sir Arthur Russell collected crystals of wulfenite from the dumps. In Hogget Gill, about a mile to the south-west, are the remains of an old, possibly 16th or 17th century, lead smelter in the walls of which occur blocks of quartz veinstone containing appreciable quantities of the iron antimony sulphide berthierite. Quartz veins with small amounts of arsenopyrite and jamesonite crop out in the adjacent stream.

Perhaps the Lake District's most famous mine is the graphite mine at Seathwaite, near the head of Borrowdale. Also known as the Plumbago, or Wad, mine, from two early names for graphite, this worked the unique deposit of graphite within a dolerite intrusion in the Borrowdale Volcanic Group. Organised mining can be traced back to the 17th century when the remarkably pure graphite was a strategically important mineral to the armaments industry. It was used for making and glazing crucibles, in casting cannon-balls and as a rust-proofing on weapons. It was also used in dying cloth, marking sheep and apparently for a variety of medical purposes. Graphite, believed to have been mined here, is known to have been employed in making moulds for counterfeit coinage. So valuable was the graphite that the miners were stripped and searched before leaving work and an armed guard was maintained at the mine. Although graphite, perhaps including some obtained here, is known to have been used as a writing medium as early as the 15th century, the beginning of pencil making, using Seathwaite graphite, did not start until the end of the 18th or beginning of the 19th century. By the early years of the 19th century several pencil mills were operating in and around Keswick, all using Seathwaite graphite. Mining at Seathwaite ended towards the end of the 19th century. The black spoil heaps from the workings are still conspicuous on the hillside above Seathwaite.

The Borrowdale Volcanic Group rocks of the Coniston area contain numerous veins, most of which

ABOVE: SEATHWAITE GRAPHITE MINE, SEATHWAITE, CUMBRIA

The dumps from levels driven into this unusual deposit are conspicuous on the hillside above Seathwaite. (©NHMPL)

MIDDLE: CONISTON COPPER MINES, CUMBRIA

Coniston Copper mines, Cumbria. The white building, now Coniston Copper Mines Youth Hostel is the former Bonsor mine office, adjacent to which parts of the extensive

dumps from this mine can be seen. Dumps from the Paddy End and other workings may be seen on the hillside in the left of this view. (B. Young)

BELOW: CUT SECTION OF BONSOR VEIN, CONISTON, CUMBRIA

In this specimen white quartz, yellow metallic chalcopyrite and black magnetite form crude bands between dark grey volcanic wall-rock. This specimen is 30 cm across. (©NERC)

are dominated by copper minerals. Tradition has it that the Romans worked copper at Coniston, although there is no reliable evidence to support this. The earliest documented mining here seems to have been by German miners under the Company of Mines Royal. Large Elizabethan surface and underground workings, including the prominent gully known as Simon's Nick, are especially conspicuous near Levers Water where, in addition, the remains of several contemporary stone dressing huts may be seen. Such was the richness of the veins that it is said that Hochstetter proposed draining Levers Water to secure further ore reserves. Although worked intermittently throughout the 17th and 18th centuries, large-scale development came in the 19th century with the principal mines at Paddy End and Bonsor. By the time the mines closed at the end of the 19th century, due to the collapse in world metal prices, the Bonsor mine had been worked to a depth of 205 fathoms (375 m) below the adit level, making it one of the deepest workings in the Lake District. In 1912 a short-lived and unsuccessful attempt was made to recover copper from the extensive dumps by an electrolytic process. Attempts to revive mining as recently as the 1950s also proved abortive. Typically the veins are composed of broken country rock and quartz in which chalcopyrite, locally with a little tennantite and bornite, are the main copper ore minerals. These are accompanied by variable concentrations of arsenopyrite, pyrite, pyrrhotite, dolomite, stilpnomelane, chlorite and traces of bismuth minerals. Small concentrations of the cobalt and nickel minerals rammelsbergite, safflorite, skutterudite and nickeline, were mined locally, together with some galena and sphalerite. A feature of the Bonsor vein, and one which contributed to its demise, was the increasing abundance of magnetite in the ore with depth. This mineral could not easily be separated from chalcopyrite with the gravity methods then available and large quantities of such ore were consigned to the dumps.

Extensive dumps remain at most of the Coniston mines, especially at Paddy End and Bonsor where, in addition, several buildings remain from the 19th century operations. The mine office is now the Copper Mines Youth Hostel; other buildings have been restored as holiday cottages. Many of the adits, including the Bonsor Deep Level, remain open, and open stopes, wheel pits, water-courses and the beds of inclined tramways, can still be seen. The Ruskin Museum at Coniston includes displays on the local mining history.

The Northern Pennine Orefield –

As has been described above, the northern Pennine orefield may be considered in two separate parts which, although sharing many common geological and mineralogical features, also exhibit important differences. In the following pages the mines and sites of mineralogical interest in the Alston and Askrigg Blocks are described separately. Many hundreds of mines and trial-workings are known and it is only possible to show a selection of these on the site maps, and to make brief comments on features of interest of some of the most significant.

The Alston Block

The date of the earliest mining in this area is unknown, although it has been suggested that the Brigantes, the Iron Age tribe that inhabited this area, may have worked iron, lead or silver. Whereas the Romans are known to have mined lead in the Askrigg Pennines, there is no definite proof that they did so here, although they occupied the area for a considerable period. Attention has also been drawn to how their most famous frontier, Hadrian's Wall, coincides in a striking way with the northern limit of workable lead mineralisation in England.

There are grounds for supposing that lead, and perhaps some silver, was being worked in the area during the Dark Ages, but it is not until the days of the Normans that we find documentary evidence of mining. A charter from King Stephen granted the right to mine for silver to his nephew, Hugh de Puiset, Bishop of Durham. Much of upper Weardale and the associated mineral rights belonged to the Bishops of Durham, perhaps from as early as 1072, and throughout medieval times much of Weardale was the Bishops' hunting forest, with a hunting park being enclosed in the 13th century. Today substantial mineral rights here are still owned by their successors, the Church Commissioners for England.

By the late 14th century, village and mine names familiar today began to appear in documents: e.g. 'Blackden' (Blackdene), 'Dawtrysheles' (Daddryshield), 'Rykhope' (Rookhope) and 'Aldwoodeclough' (Allercleugh). There is also evidence, from legal documents relating to a dispute between W. Blackett and Isaac Basire, that most of the Weardale deposits had been discovered by 1666. Several northern Pennine vein and place names, e.g. Henrake (Heinrich), Rowpotts (Rupert's) veins and Blagill (Bleigill), may date from the introduction of German miners to the area by Elizabeth I.

The estates of the Earl of Derwentwater, including the valuable mineral rights of Alston Moor, were forfeited after the 1715 rebellion and passed to the Commissioners of Greenwich Hospital. The huge expansion of mining in the 18th and 19th centuries proved extremely profitable to these new owners. These developments in mining and smelting, which began in the 18th century, were to establish the area as a major mining field for the next 200 years. Although, as we shall see, two major concerns dominated the industry in this part of the Pennines, about a third of lead produced came from small companies in the 19th century.

The first of the two largest concerns was a Quaker company founded in 1692, formally known as the 'Governor and Company for Smelting Down Lead with Pit Coal and Sea Coal', better known as the 'London Lead Company'. This company merged in 1705 with another Quaker-owned company, the Ryton Company, that originally had taken leases in Alston Moor in 1696. During the 18th century the London Lead Company acquired more leases and extended its interest into Teesdale and the Derwent Valley. In 1801 the company leased a number of mines in Weardale and built a smelt mill at Stanhopeburn.

The second was a local family business begun by Sir William Blackett (1621-80), an extremely wealthy Newcastle merchant who had widespread commercial interests, including coal and lead mining. The exact date of his first venture into lead mining is not known, but he mined in Allendale and Weardale and on his own estates in the Tyne Valley. Although the family

name changed to Beaumont through marriage, the business remained a family-run concern. Known throughout its life as W. B. Lead, this business owned and operated mines and smelt mills in the Allendales and Weardale.

Under these two large enterprises, the veins were systematically explored and mines planned and operated in the most efficient ways possible. Accurate plans were kept of all workings, and production and grades of ore were recorded. Because they held leases over

1	Settlingstones mine	44	Copt Hill Quarry
2	Stonecroft & Greyside mines	45	Burtree Pasture mine
3	Fallowfield mine	46	Sedling mine
4	Langley Barony mine	47	Ireshopeburn or Barbary mine
5	Cockmount Hill mine	48	Blackdene mine
6	Langley Smelt Mill	49	Levelgate mine
7	Dukesfield Smelt Mill	50	Old Fall mine
8	Allen Smelt Mill	51	Middlehope mine
9	Tindale Fell Smelt Mill	52	Slitt mine
10	Longcleugh mine	53	West Rigg Quarry
11	Ouston mine	54	Heights Quarry and mine
12	Sipton mine	55	Cambokeels (Cammock Eals) mine
13	St Peter's mine	56	Billing Hills mine
14	Allenheads Smelt Mill	57	Eastgate Quarry
15	Allenheads or Beaumont mine	58	Westernhope mine
16	Swinhopehead mine	59	Greenlaws mine
17	Acton Smelt Mill	60	Rowantree mine
18	Shildon mines	61	Carricks mine
19	Reeding mine	62	Harnisha Burn mine
20	Silvertongue mine	63	Yew Tree mine
21	Healeyfield mine	64	Parson Byers Quarry
22	Swandale, Burnhope and Harehope Gill mines	65	Cornish Hush
23	Jeffrey's mine	66	Whitfield Brow mine
24	Ramshaw mine	67	Harehope Gill mine
25	Whiteheaps mine	68	Hartside mine
26	Sikehead mine	69	Ardale Head mine
27	Frazer's Hush & Greencleugh mines	70	Cash Burn mine
28	Groverake mine	71	Cornriggs mine
29	Wolfcleugh mine	72	Cashwell mine
30	Rispey mine	73	Silverband mine
31	Rookhope or Lintzgarth Smelt Mill	74	Swathbeck mine
32	Redburn mine	75	Loppysike mine
33	Boltsburn mine	76	Dun Fell mine
34	Stotfield Burn mine	77	Dufton mine
35	Brandon Walls mine	78	Calvert mine
36	Stanhopeburn Smelt Mill	79	Tynehead Smelt Mill
37	Stanhopeburn mine	80	Tyne Green mines
38	West Pasture mine	81	Greenhurth mine
39	Noah's Ark Quarry	82	Dubbysike mine
40	Hope Level mine	83	Cowgreen mine
41	Rogerley Quarry	84	Willyhole & Reddycombe mines
42	Drypry mine	85	Lady's Rake mine
43	Park Level and Killhope Lead Mining Museum	86	Ashgill Head mine
		87	Grasshill mines
		88	Trough mine
		89	Bands mine

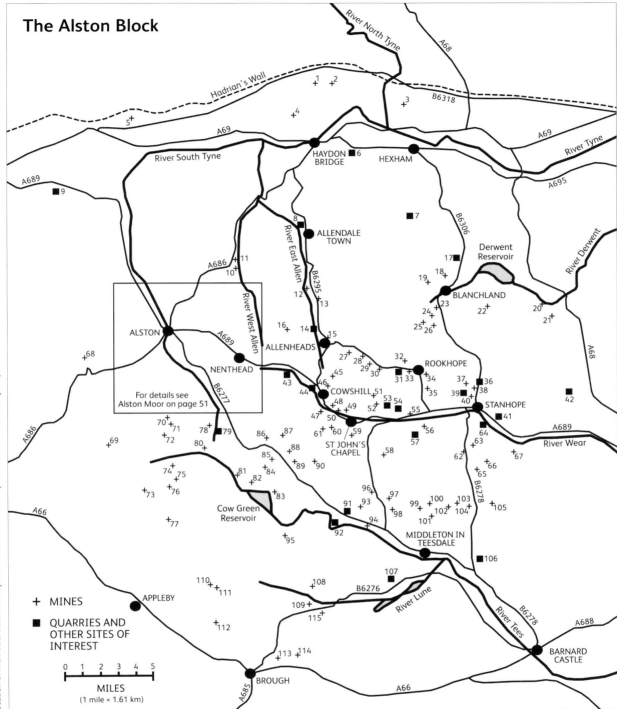

The Alston Block

For reasons of clarity, only a selection of roads, rivers and settlements are shown. For more detailed location of individual sites this map should be used with an up-to-date Ordnance Survey map.

River North Tyne

Hadrian's Wall

B6318

A68

A69

River Tyne

A695

River South Tyne

HAYDON BRIDGE

HEXHAM

A689

River East Allen

ALLENDALE TOWN

B6306

Derwent Reservoir

River Derwent

A686

River West Allen

ALSTON

A689

NENTHEAD

B6277

For details see Alston Moor on page 51

ALLENHEADS

BLANCHLAND

A68

COWSHILL

ROOKHOPE

STANHOPE

River Wear

A689

ST JOHN'S CHAPEL

MIDDLETON IN TEESDALE

A686

A66

Cow Green Reservoir

B6278

B6276

River Lune

River Tees

B6278

A688

A66

BARNARD CASTLE

APPLEBY

+ MINES

■ QUARRIES AND OTHER SITES OF INTEREST

BROUGH

A685

0 1 2 3 4 5

MILES
(1 mile = 1.61 km)

90 Langdon mines	96 Pike Law mines	103 Manorgill mine	110 Murton mine
91 High Force Quarry	97 Flushiemere mine	104 California mine	111 Hilton mine
92 Force Garth (Middleton) Quarry	98 Red Grooves mine	105 Sharnberry mine	112 Long Fell mine
93 Ettersgill mine	99 Coldberry mine	106 Blackton Smelt Mill	113 Augill mine
94 Wynch Bridge mines	100 Lodgesike mine	107 Greengates Quarry	114 Cabbish mine
95 Yorkshire Silverband mine	101 Skears mine	108 Closehouse mine	115 Forster's Hush
	102 Wiregill mine	109 Lunehead mine	

45

large areas, they could co-ordinate exploration, mine development, haulage and drainage using well-planned levels and shafts, which in some cases served several originally separate mines. Notable among these enterprises was the driving of long drainage adits, which also explored for hitherto unknown deposits, such as the 5 mile (8 km) Nentforce Level from Alston to Nenthead, and the 7 mile (11.3 km) Blackett Level, planned to run from Allendale Town to Allenheads but discontinued after 4.7 miles (7.5 km).

Improvements in mining and ore dressing were matched by developments in smelting. Both companies operated large and efficient smelt mills designed to treat the produce of several mines. Significant innovations in smelting technology were made in these mills. Perhaps best known of these was the crystallisation process for improving the efficiency of silver recovery from newly-smelted lead, devised by Hugh Lee Pattinson at W. B. Lead's Blaydon refinery and known as the Pattinson process.

Silver was always a significant by-product of lead smelting in the area, as virtually all of the orefield's galena contains silver. During the hey-day of mining and smelting in the 18th and 19th centuries, silver recovery values averaged about 3.5 oz of silver per ton of lead (approximately 100 ppm silver), although assay values of up to 90 oz of silver per ton of lead (approximately 2500 ppm silver) were recorded from a few localities. It has been suggested that 12th-century references to the 'Carlisle Silver Mines' may relate, at least in part, to the mines of Alston Moor. It is known that the Bishop of Durham operated a mint in Durham in the late 12th century, and it seems likely that north Pennine silver, probably from Weardale, or Shildon in the Derwent Valley, may have contributed to the wealth of the Bishops of Durham at about this time. If unsubstantiated historical claims of very high silver outputs from the area in the 11th and 12th centuries are correct, ores very much richer in silver than any known today, or recorded during the peak years of mining and smelting in the 18th and 19th centuries, must then have been available. If such enriched silver ores occurred, no traces of them have survived. The wealth of reliable records of silver assay and recovery values, gathered from across the orefield, do not support the claims of some economic historians for very high levels of silver production in the medieval period, and recent historical research suggests that at least some of the claims for early high silver outputs may have been greatly exaggerated.

Whatever the reality of medieval silver production, such was the output of silver supplied to the Mint from the London Lead Company's smelt mills in the 18th century, much of it from the northern Pennines, that the Company requested that its origin be acknowledged by an appropriate 'device' on the coinage of the day.

PATTINSON PANS

Pattinson pans, used in the separation of lead and silver during refining, at Nenthead Smelt Mill, Nenthead, Cumbria. In this process, lead was melted in a series of pans, each kept at a crucial temperature by a small fire. On cooling slowly, lead crystals began to form which Pattinson found were poor in silver compared to the original melt. By repeating the process, the remaining liquid metal became enriched in silver, which was then recovered by the cupellation process. Although other techniques for silver extraction were used, it was the Pattinson process that continued to be the most favoured in the northern Pennines. (Beamish)

Old postcard showing a view across Nenthead village in the early 20th century, in the centre of which may be seen the large plant built to treat both lead and zinc ores by the Vielles Montagne Zinc Co. in 1908. Parts of this building are still in use as a bus garage.

The request was granted, and most of the five-shilling, two-shilling and sixpenny pieces minted between 1706 and 1737 (i.e. on coins of Anne, George I and George II) carried the device of 'Two roses and feathers quartered as the emblems of England and Wales' to denote the domestic sources of the silver. These coins were long known as 'Quaker Shillings'.

The influence of the two major mining companies was not confined to the technology of mining and smelting. Both took an active interest in their local communities. Much of the area's infrastructure today can be related directly to the activities of the mining companies. Several of the dales villages were built or greatly enlarged by them: e.g. Allenheads and Carrshield by W. B. Lead; Nenthead by the London Lead Company. Other settlements were also substantially altered or enlarged: e.g. Middleton-in-Teesdale by the London Lead Company who established their northern Pennine headquarters here. Not only houses but schools, churches, public houses, reading rooms and libraries were provided. Welfare schemes, social amenities and the education of their workers and families were priorities of both businesses. It was claimed that some of the libraries were more popular with the miners than the public houses. Perhaps this was just as well, as drunkenness was usually rewarded by dismissal. Great importance was attached to education, particularly

science and engineering, and the most promising students of all ages were encouraged to go on to further study.

A traditional feature of northern dales life was that of the miner-farmer. Many miners held small holdings as an extra means of support, and as a refreshing respite from heavy underground labour. More cynically from the company's point of view, this relationship helped establish a close attachment to the land, acting as a disincentive for the miner to move away to other employment. It was customary to find work in the mines suspended briefly at critical times of the year such as at hay-making. The pattern of enclosed fields high on the fellsides with scattered farm houses and barns, many now derelict, is one of the most distinctive features of the northern Pennine landscape.

In an upland area of this sort, many mines were remote even from upland small holdings. It was usual at such mines for the company to provide a lodging house or 'shop' at which men stayed during the week. Accommodation in such places was, to say the least, basic. A government commissioner, appointed to look into the conditions of children working in the mines in the 19th century, famously commented that he would rather spend 24 hours locked in a mine than 15 minutes in a mine 'shop'! Numerous deserted shops stand close to mine entrances high in the dales as gaunt reminders of this long vanished way of life.

Remoteness also posed the problems of bringing materials to the mines, and of transporting ore and finished lead away from the mines and smelt mills. The solution adopted by the companies, and in some cases the owners of the mineral rights, was to develop the road network which, apart from a few modern additions, is substantially that seen today.

This then was the orefield in its better days. Towards the end of the 19th century the combination of a collapse in world metal prices, and rising costs of production, proved disastrous for most mines. W. B. Lead surrendered its leases in Weardale in 1880 and the London Lead Company ceased production and was finally wound up in 1905. The Weardale leases were taken over by the newly formed Weardale Lead Company in 1883. After some poor times, these were to become profitable with the success of mines such as Boltsburn. During the second half of the 20th century several mines of the Weardale Lead Company were eventually taken over by a number of other operators including ICI (Imperial Chemical Industries), Swiss Aluminium Mining, British Steel Corporation, Minworth and Weardale Fluorspar, all to produce fluorspar.

Zinc mining, started by the London Lead Company in the Alston area, to some extent replaced lead mining with the establishment of the Nenthead and Tynedale Lead and Zinc Company, who were later succeeded in 1896 by the Vieille Montagne Zinc Company of Liege, Belgium, under whose tenure zinc production continued until the late 1930s.

It is often forgotten, or not realised, that the northern Pennines also produced important quantities of iron ores. Indeed, one of the earliest references to mining in the area, dating from the 12th century, relates to the granting of a lease for an iron mine at Rookhope. Siderite and ankerite, and their weathering product 'limonite', are important constituents of many vein and flat deposits. The opening up of substantial deposits by the Weardale Iron Company, formed in 1842 by Charles Attwood (1792-1875), created employment for 1700 men by the late 1850s. Attwood used the Bessemer process, patented in 1856, to produce steel from local ores at the Wolsingham Iron Works, which he established in 1864. There were also blast furnaces at Stanhope. By the end of the 19th century the iron industry was in decline and significant iron mining ended early in the 20th century, although there was a brief resumption of mining and exploration during the Second World War.

The development of mining for the non-metallic minerals fluorite, barite and witherite, also helped to offset the economic decline brought about by the demise of lead mining. Witherite mining began in the Alston area as early as 1850, and over the next century this mineral became a major product of the orefield from several mines, most notably Settlingstones in the Tyne Valley, which maintained production until 1969. Witherite was also produced from several collieries in the Durham coalfield where witherite veins cut Coal Measures rocks. The abundance of witherite is a unique feature of this orefield which, for long periods, was the world's sole commercial source of this rare industrial mineral. Barite mining was important, particularly in the early years of the 20th century, and continued until the early 1990s.

Fluorspar mining was started by the Weardale Iron Company in 1882 and was rapidly developed as an important part of the local economy by the Weardale Lead Company, in part as a by-product of lead mining. It subsequently became the main mineral sought. Such was the demand that abandoned mines were reopened and old dumps reprocessed to obtain it. By the 1990s competition from cheap Chinese fluorspar had rendered the remaining northern Pennine mines uneconomic, and with the closure of the combined Frazer's Hush-Groverake mine in 1999 large-scale mining in the northern Pennines came to an end.

The Pennine Escarpment

The steep, high escarpment of the northern Pennines, which extends south-east from Hartside Pass to Brough, includes a handful of mines, most of which lie within the outer, barium, zone of the orefield.

Remotely situated at Hard Rigg Edge on Melmerby Fell are the remains of a working of unknown, but probably early, date. The small dumps have yielded some good specimens of deep green pyromorphite crystals, a rather uncommon mineral in this orefield. Interesting quartz pseudomorphs after 'cock's-comb' barite have also been found.

LEFT: FLAT DEPOSITS, DOW SCAR HIGH LEVEL, HILTON MINES, SCORDALE, CUMBRIA

Bob Symes examines irregular bands of coarsely crystalline white barite in fluoritised Melmerby Scar Limestone exposed in the underground workings. (©NHMPL)

ABOVE: YELLOW FLUORITE, SCORDALE

Group of bright amber-yellow fluorite cubes, typical of those from the flat deposits at Hilton mines, Scordale, Cumbria. In this photograph of NHM spec. no. BM 1964R. 1946, the field of view is approximately 7 cm. The specimen is part of the Russell Collection. (©NHMPL)

Above the village of Knock, and almost at the summit of Great Dun Fell, lies the site of the Silverband mine (better referred to as 'Westmorland Silverband' to avoid confusion with the 'Yorkshire Silverband' mine of Teesdale). Originally worked in the 19th century as a lead mine by the London Lead Company, the abundance of white barite in extensive flats in the Great Limestone led to its reworking for this mineral by the Laporte Company between 1939 and 1963. A further period of reworking, also for barite and this time by opencast methods, took place during the 1970s and '80s. The flats at Silverband mine have yielded some magnificent specimens of crystallised barite, both as 'cock's-comb' aggregates and as large, stout water-clear tabular crystals.

About three miles south-east of Silverband mine, and also high on the escarpment, are the remains of the Dufton mines. Veins and flats, here in the Tynebottom, Jew and Smiddy limestones, were mined for lead ore in the 19th century by the London Lead Company. In 1882 Dufton Fell mine reopened to produce barite, one of the first mines in the area to do so, and continued in operation until 1897. The dumps were treated to recover the remaining barite during the 1980s. Dufton mines were celebrated for very fine crystals of colour-

less barite. Some were especially large, including one weighing over 500 kg, now in the collection of the Natural History Museum, London. Very fine examples of 'flos-ferri' aragonite were also found in the Dufton flats.

At the head of the steep-sided valley of Scordale, which cuts the Pennine escarpment above the village of Hilton, a cluster of roughly E-W trending veins and associated flats in the Melmerby Scar Limestone were worked at the Murton and Hilton mines. These lie respectively west and east of Hilton Beck. Both groups of mines were developed as lead producers early in the 19th century by the London Lead Company, who worked them until 1876. They were reopened for barite production in 1896 and, under the Scordale Mining Company, and later the Brough Barytes Company, mining continued until 1919. A little witherite was also mined at this time from a small flat in the Murton mines known as the 'Carbonate Shake'. The mineralisation at the Scordale mines is rather unusual for the northern Pennines, as both fluorite and barite are present in abundance. Fluorite typically occurs as pale to deep yellow cubes, commonly overgrown by large tabular white barite crystals. The mines, especially the Hilton mines, are justifiably famous for superb deep yellow or amber cubic crystals collected from the many cavities in the flats. A pocket of nickel-rich ore, consisting of nickeline and gersdorffite with traces of millerite and some superbly crystallised supergene annabergite, has been described from the Dow Scar vein in the Hilton mines. These mines are within the Warcop military training area and access is severely limited.

Alston Moor

Alston Moor, the remote and lonely upland lying south of the small market town of Alston, and including the valleys of the Nent, Black Burn and upper reaches of the South Tyne, is a name familiar to most mineralogists and collectors. Not only does the area have a long and distinguished history of mining, but it has been a prolific source of many beautiful mineral specimens.

One of the greatest concentrations of deposits within this area occurs in the valley of the River Nent, which joins the River South Tyne at Alston. The valley lies mainly within the outer, barium-rich, zone of the orefield, although many deposits around Nenthead include neither barium minerals nor fluorite.

Prominent on the north side of the Nent Valley are the dumps and surface workings of Blagill mine. The Alston Moor mines are said to have been worked in the 14th century by German miners under one Tilman of Cologne, and it has been suggested that the name 'Blagill' may be derived from the German 'blei', meaning lead. Working at Blagill may date back at least that far, although it was the 18th century that saw the most profitable mining here. Originally worked for lead ore, the principal vein, the NE-SW trending Fistas Rake, carried an abundance of the rare barium calcium carbonate mineral barytocalcite for which Blagill is the type locality. Between 1876 and 1895 this mineral was mined here commercially as a low grade form of witherite, but the discovery of abundant true witherite at the nearby Nentsberry mine at about this time effectively destroyed the market for the Blagill mineral. Spectacular specimens of barytocalcite, collected last century, are to be found in most museum collections, and even today good specimens may still be seen at

FLUORITE, ALSTON, CUMBRIA

Clear deep purple interpenetrant twinned crystals of fluorite on quartz. This specimen, c.6.5 cm across [NHM spec. no. BM 44507], labelled 'Alston Moor, Cumberland', was purchased from William Burrows in December 1871. Fine specimens of this sort, also with similarly imprecise locality details, are characteristic of Alston Moor and Weardale. (©NHMPL)

BLAGILL MINE SPOIL HEAPS

The spoil heaps of Blagill Mine, near Alston, Cumbria. The site is today scheduled as a Site of Special Scientific Interest (SSSI) for its abundance of barytocalcite, for which it is the type locality. (©NHMPL)

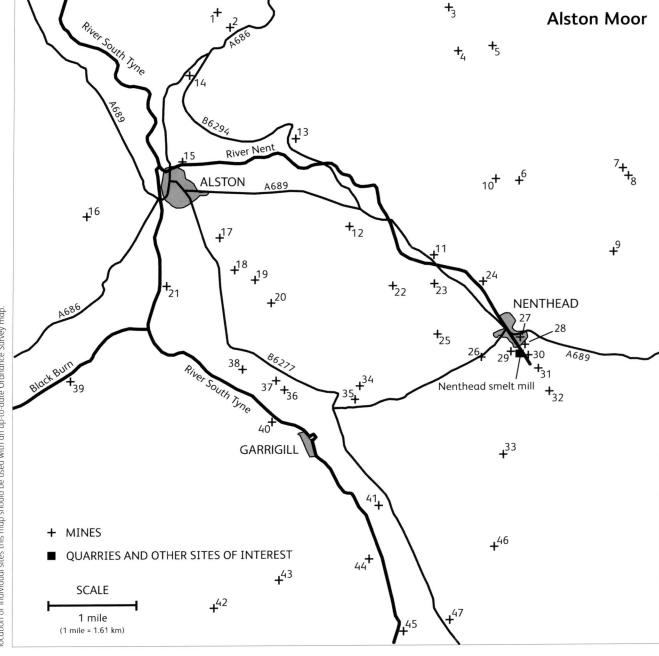

Alston Moor

For reasons of clarity, only a selection of roads, rivers and settlements are shown. For more detailed location of individual sites this map should be used with an up-to-date Ordnance Survey map.

+ MINES

■ QUARRIES AND OTHER SITES OF INTEREST

SCALE

1 mile
(1 mile = 1.61 km)

1	Ayle Burn mine	13	Blagill mine
2	Clargill mine	14	Taylor's Grove mine
3	Mohopehead mine	15	Nentforce Level
4	Heartycleugh mine	16	Park mines
5	Hesleywell Hush	17	Bayle Hill mine
6	Wellhopehead mine	18	Farnberry & Holyfield mines
7	Scraithole mine	19	Nattrass mines
8	Barneycraig mine	20	Dowpot Sike mine
9	Coalcleugh mine	21	Nest mines
10	Wellhope Shaft	22	Galligill Sike mine
11	Nentsberry Haggs mine	23	Grassfield mine
12	Hudgill Burn mine	24	Brownley Hill mine

25	Greengill mine	37	High Hundybridge mine
26	Dowgang mine	38	Crag Green mine
27	Brewery Shaft	39	Rotherhope (Rodderup) Fell mine
28	Rampgill mine	40	Tynebottom mine
29	Capelcleugh mine	41	Ashgill Field mine
30	Carr's mine	42	Pikeman Hill Quarry
31	Smallcleugh mine	43	Crossgill Copper mines
32	Middlecleugh mine	44	Leehouse Well mine
33	Wellhopeknot mine	45	Sir John's mine
34	Bentyfield mine	46	Little Gill mine
35	Whitesike mine	47	Windy Brow mine
36	Cowper Dyke Heads mine		

51

Blagill and a few other nearby mines. Today the site is protected as a Site of Special Scientific Interest because of the abundance of this mineral.

About two miles up the valley from Blagill, the modest gated entrance of Nentsberry Haggs Horse Level belies the true extent of this long-famous mine, the most extensive and recent workings of which lie over a mile to the north beyond the Cumbrian boundary in Northumberland. The level, and the shallow workings on the hillside above it, are on the outcrop of the Nentsberry Haggs vein from which lead ore was worked. However, the level was continued through a number of interconnecting veins, eventually reaching a complex of veins beneath the Wellhope Valley. These proved to contain substantial reserves of galena, sphalerite and witherite. Rather than accessing these via the rather tortuous Nentsberry Haggs Level, in 1925 the operators, the Vieille Montagne Zinc Company, sank the new Wellhope Shaft. It was planned to carry crude ore from here to the treatment plant in Nenthead on a specially built aerial ropeway, although as this never worked satisfactorily it was soon demolished. Some of the foundations for its towers may still be seen. Difficulty was experienced in treating the mixed sphalerite-witherite ore and substantial amounts of this are believed to remain *in situ* underground. Mining ended in 1938 and, despite several attempts at reopening, the mine has since lain idle. Nentsberry Haggs mine has left a substantial legacy of fine specimens, including excellent witherite crystals, commonly with white surface alteration to barite, sphalerite and some magnificent examples of barytocalcite. The latter includes clusters of beautiful honey-coloured crystals collected from the Second Sun vein, and very distinctive lenticular crystals, typically coated in barite and calcite, found in a series of replacement deposits in the Great Limestone known as the Admiralty Flats. Within the last few years the mine has also yielded some excellent alstonite crystals, many partly or wholly pseudomorphed by calcite or barite.

Another of Alston Moor's famous mines, Brownley Hill mine, lies on the north side of the valley about a mile upstream from Nentsberry. Like most of the mines in this area, Brownley Hill worked a number of veins and associated flats. Working may be traced back to at least the 18th century when, for a period, the London

Lead Company held the mine. In common with many of the Nenthead deposits, those at Brownley Hill contained an abundance of sphalerite as well as galena, and from as early as the 19th century both lead and zinc ores were worked by a variety of companies. Mining ended in 1896, although for some years the mine was used as a haulage way for ore from Nentsberry Haggs mine, with which it is connected. An attempt to reopen the mine for zinc ore by the Vieille Montagne Company in 1936 was abandoned after only six months.

Brownley Hill mine is best known to mineralogists as a source of the rare barium calcium carbonate mineral alstonite. It is often regarded as the type locality for this species, although the original descriptions of the mineral relate to specimens from both here and Fallowfield mine, near Hexham. When first discovered, this mineral was regarded as a form of barytocalcite and was originally referred to as 'right rhombic barytocalcite'. The name 'bromlite', based on a mis-spelling of 'Brownley', was proposed shortly afterwards by Thomson (1837), and in 1841 Breithaupt introduced the name 'alstonite' which thereafter became the accepted name, despite the apparent priority of bromlite. Almost since its discovery alstonite seems to have been sought after by collectors; Nall, writing in 1888, commented that the parts of the mine from which it was obtained were by then inaccessible and that even the smallest specimens were then fetching as much as five pounds. Underground exploration by collectors during the 1980s traced the source of the original specimens to a comparatively restricted part of the High Cross vein, and not Jug vein as was previously supposed. In 1993 the mine became the type locality for a second new mineral. Brianyoungite, a zinc carbonate sulphate hydrate, was found in association with gypsum in oxidised parts of the Wellgill Cross vein.

The main entrance to Brownley Hill mine was the Bloomsberry Horse Level, the stone-arched portal of which can be seen near the junction of the River Nent with Gudham Gill. This same entrance gave access to the workings of Gudhamgill mine in which a series of flats contained numerous vugs lined with yellow fluorite, sphalerite, quartz and galena.

The village of Nenthead, at the head of the Nent Valley, owes its existence to the activities of the London Lead Company. It was they who planned the

ABOVE: BRIANYOUNGITE

Tiny white spherules (up to approximately 0.1 mm across) on gypsum,
from the type locality in old underground workings in the Wellgill
Cross Vein, Brownley Hill mine,
Nenthead, Cumbria. (©NERC)

RIGHT: DOWGANG HUSH

Looking northward along Dowgang Hush, Nenthead, Cumbria,
one of the largest such opencast
workings in the area, excavated on
the course of the Dowgang vein.
(B. Young)

village as they came to recognise the great extent and
value of the lead deposits here. Much of the village
was constructed in the 1820s, centred around the row
of cottages known as Hillersdon Terrace. Despite the
long abandonment of the mines, and the demolition
of some buildings, and a few more recent additions,
significant parts of the village are substantially as built
by the company. Notable buildings which date from
this period include the church, chapel, miners' reading
room, the Miners' Arms public house and rows of
cottages. The remains of the huge dressing plant, built
by the Vieille Montagne Company in 1909, and now
used as a bus garage, still dominate the village. Lead
and zinc concentrates, produced here from a number
of local mines, were despatched to the company's plant
in Liege for smelting. During the Second World War a
new flotation plant was installed in the building to
extract many tonnes of zinc and lead concentrates for
the war effort by reprocessing the many mine dumps in
the neighbourhood. Shortly after the war, Anglo
Austral Mines, a subsidiary of the Imperial Smelting
Corporation, acquired the former Vieille Montagne
Company leases and adapted the mill to process
fluorite from mines at Rotherhope Fell and Weardale.

The deep-wooded valley of Dowgang Hush, on
the hillside south of the village, is a huge excavation,
in part worked by 'hushing', on the outcrop of the

Dowgang vein. The hush is especially spectacular
when seen from alongside the Nenthead to Garrigill
hill road.

On the south side of the village, beyond the car
park that occupies the site of the original London Lead
Company's Rampgill dressing floors, is the relatively
inconspicuous entrance to Rampgill mine. Although
discovered and worked before 1745, it was not until
after the London Lead Company obtained the lease
that the mine prospered. It is said that the ground
opened up along Rampgill vein was the richest ever
worked on Alston Moor. Towards the end of the 19th
century the mine was acquired by the Vieille Montagne
Company who extended the level north-eastwards,
discovering rich flats in the Great Limestone beneath
Coalcleugh at the head of West Allendale. Although
the Rampgill deposits lie within the fluorite zone of
the orefield, the mineral is scarce in the vein, although
quartz epimorphs after fluorite cubes were described
as common by contemporary observers. Many fine
examples of these have been obtained by collectors
who have accessed these workings in recent years.

Adjacent to the Rampgill mine entrance, the
restored stone-built mine workshops house an interpretation centre of local mining history run by the
North Pennine Heritage Trust whose headquarters are
also here. A short distance higher up the valley, beyond

RAMPGILL MINE

Miners and pony at the entrance to Rampgill Horse Level, Nenthead, Cumbria, c.1910. (Beamish)

the recently constructed water power exhibits, are the remnants of the London Lead Company's Nenthead Smelt Mill. Although most of the mill is very fragmentary, remains of its associated flues, condensers, wheelpits and other buildings still convey a clear impression of the former importance of this site. Especially impressive is the magnificent assay house with its tall chimney.

On the hillside a short distance above the Rampgill site, and connected to it by a footpath, is a small clump of trees surrounding the collar of the Brewery Shaft. This 100m-deep shaft connects with the end of the Nentforce Level, an 8km-long tunnel driven from Nentforce waterfall at Alston. This was designed both to drain the mines of the valley and to explore for undiscovered veins. Promoted by the Commissioners of Greenwich Hospital, the owners of the mineral rights, and planned by John Smeaton, the famous civil engineer, and Nicholas Walton, driving was begun in 1776, but it was not until sometime between 1842 and 1856 that it was extended to the Brewery Shaft. Like the Blackett Level in East Allendale, the Nentforce Level failed in many of its objectives. By the time it had been driven sufficiently far up the valley, the mines it was intended to drain were either abandoned or had been sunk beneath this level. Only one significant new orebody was discovered along its length on the Hudgill Burn Cross vein. However, during the Vieille Montagne Company's tenure of Rampgill mine, water from the

ingenious air compressors installed in Brewery Shaft drained away through the Nentforce Level.

The upper part of the Nent Valley, one of the most intensely mineralised parts of the orefield, is riddled with interconnecting underground workings, and huge spreads of mine spoil mantle the fellsides on both sides of the river.

Most famous of these workings is Smallcleugh mine. Although originally attracted by the Smallcleugh and Handsome Mea Cross veins, it was the London Lead Company's discovery of the extensive flats associated with these that were to bring fame to the mine. They proved to extend over a length of 1.1 km within an up-faulted, or horst, structure in the Great Limestone between the two veins. The flats consist of extremely hard ankeritised limestone in which lenses of galena and sphalerite are common. Cavities lined with ankerite, quartz, galena and sphalerite crystals are abundant. It has been estimated that at least 5 million cubic feet (141,585 cubic metres) of limestone had been altered to produce these remarkable deposits. Originally worked for lead ore alone, it was the abundance of sphalerite which was to give the mine a new lease of life in the latter years of the 19th century when reworked for zinc. During its great years, the Smallcleugh workforce was numbered in hundreds, including a large contingent of Italian miners brought to Nenthead by the Vieille Montagne Company. Their monu-

SMALLCLEUGH MINE, NENTHEAD, CUMBRIA

Bob Symes stands in the underground workings in the flats of Small-cleugh mine, Nenthead. The Great Limestone is here extensively replaced by ankerite and quartz with discontinuous bands and pockets of galena and sphalerite. Remnants of bedding in the original limestone, and several large vugs, are seen in the right of the view. (©NHMPL)

ment is the extensive series of huge underground chambers, including the famous 'Ballroom flat' in which in 1901 the local Masonic lodge held a formal dinner. It is difficult when visiting this lonely site today, either on the surface or underground, to imagine that only a little over a century ago this was an important centre of industry.

To the mineralogist Smallcleugh mine is well-known as a source of beautifully crystallised black sphalerite, rather dull large galena crystals, and pale cream anker-ite. Parts of the old workings are notable for super-gene minerals including delicate acicular epsomite crystals, turquoise-blue melanterite and small speci-mens of serpierite and namuwite.

Between Smallcleugh mine and the smelt mill is the recently restored entrance to Carr's mine, parts of the underground workings of which are now open to visitors to the Nenthead Mines Heritage Centre.

Before leaving the Nent Valley, one other mine calls for comment. Hudgill Burn mine lies on the south side of the valley between Nenthead and Alston. Today the site is a caravan park adjacent to a large tree-covered spoil heap. Trials near here by the London Lead Com-pany and others late in the 18th century met with no success. It fell to the local partnership of John and Jacob Wilson and Company to discover the remark-able group of veins which, over the course of the next 60 years or so, were to prove some of the richest on Alston Moor. The Hudgill Burn deposits were unusual in containing an abundance of cerussite as well as galena. So altered was the ore that it could be worked by hand without resort to blasting. Massive hydrozin-cite was also found. Perhaps rather surprisingly, very few specimens from here are to be found in collections.

Deeply oxidised ores, similar to those at Hudgill Burn, were found in several mines of Middle Fell, the ridge which lies between the Nent and South Tyne valleys south-east of Alston. Deposits of smithsonite, known locally as 'calamine', were worked at Bayle Hill, Farnberry and Holyfield mines. Most of this mineral occurs here as the brown porous, cellular 'dry bone' variety, but at the Farnberry and Holyfield mines beautiful stalagmitic masses of compact yellowish green crystalline smithsonite were common.

Of the numerous mines of the upper South Tyne Valley, perhaps best known to mineralogists is Tyne-bottom mine, alongside the Pennine Way on the south bank of the river immediately west of Garrigill. This comparatively small mine appears to have been first worked by the Earl of Carlisle in the latter part of the 18th century, before being taken by the London Lead Company who worked it until 1873. The veins here are associated with flats in the Tynebottom Limestone. Apart from galena, for which the mine was worked, the mineralisation includes quartz, purple fluorite and calcite, the latter commonly in a variety of crystal morphologies. Many very fine specimens of calcite were obtained here and it is likely that many of the numerous well-crystallised calcite specimens in old collections, simply labelled 'Alston Moor', may have originated at Tynebottom. The mine is also famous for the abundance of vivid pink crusts of erythrite, formed by post-mining oxidation of cobalt minerals such as glaucodot which occurs here as microscopic inclusions in pyrite and marcasite. Other supergene minerals for which the mine is noted include devilline, serpierite, brochantite and wroewolfeite. An assem-blage of silver minerals, including argentopyrite, sternbergite, pyrargyrite, stephanite and acanthite, has been described from here, based on specimens said to have been collected here by A. W. G. Kingsbury. In common with many of Kingsbury's more remarkable 'finds', these specimens almost certainly did not originate here.

South of Garrigill are several small mines, some of which were worked principally for copper, although only very small amounts of ore were obtained. Sir John's mine, on the east bank of the South Tyne, which worked lead from the Sir John's vein, also cut the Great Sulphur vein. This powerful vein, occasionally referred to by the Alston Moor miners as the 'Backbone of the Earth', is unique in the orefield for its great width, usually many metres wide, and its composition, being dominated by quartz with pyrite, pyrrhotite and marcasite. A little chalcopyrite occurs locally, but galena and sphalerite are very rare. Despite its great size, the vein has yielded little of economic value, although it was investigated as a source of pyrite for sulphur-making during the Second World War. An earlier, but entirely fruitless attempt, was made to extract gold from the vein at Sir John's mine.

One of Alston Moor's most recently-worked mines, Rotherhope Fell mine, lies in the valley of the Black Burn, about three miles south-south-west of Alston. Although the first workings probably date from the 18th century, the main period of activity was throughout the 19th and first half of the 20th century. During this time the original Rodderup Fell Lead Mining Company was succeeded by the Vieille Montagne Zinc Company, who in turn were followed by Anglo Austral Mines. The main working level was engineered by John Smeaton, whose involvement in the Nentforce Level has already been noted. The veins were associated with large flats which, like those at Tynebottom mine, were developed in the Tynebottom Limestone. Originally worked for lead, the mine became a significant fluorspar producer in the 20th century. Although underground mining ended in 1948, some fluorspar was recovered from the spoil heaps during the 1970s. Rotherhope Fell mine's main claim to mineralogical distinction lies in the beautiful purple fluorite cubes with rich yellow margins, discovered in a small area in the flats.

West Allendale

A small cluster of ENE-WSW trending veins has been worked for lead ore in the Great Limestone near the village of Ninebanks, mainly in the 19th century. Although extensive exploration was undertaken on some of these by W. B. Lead, records indicate that the amount of ore obtained was modest. Grassed-over spoil heaps and overgrown hushes are especially conspicuous on the Keirsleywell Row veins near Ninebanks Youth Hostel. Unsuccessful attempts were made to locate workable reserves of witherite at the nearby Longcleugh mine in the 1970s.

The rather meagre remains of the Mohopehead and Heartycleugh mines may be seen in the tributary valley of the Mohope Burn, south of Ninebanks. In common with most of the veins of West Allendale, those worked here contained abundant barium minerals, both barite and witherite. Barytocalcite was abundant at Heartycleugh mine and has also been found in smaller amount at Mohopehead, Longcleugh and other mines in the valley. Workings, remotely situated at Wellhope, at the head of the Mohope Valley, explored parts of the complex of veins extensively worked from Nentsberry mine in the Nent Valley, described earlier under the heading of Alston Moor.

The most conspicuous evidence of mining in West Allendale today may be seen around the hamlets of Carrshield and Coalcleugh. Although Carrshield is a tiny village today, some impression of its importance in the hey-day of 19th-century mining may be gained from the impressively large village school, built by W. B. Lead and now converted to private houses. Large spoil heaps and a dilapidated mine shop to the south of the village mark the sites of the Scraithole and Barneycraig mines.

At Scraithole mine W. B. Lead's 19th-century exploration of a long length of Scraithole vein was rewarded with only small amounts of lead ore. However, the vein, and associated flats, were found to contain an abundance of sphalerite and witherite, similar to the mineralisation found in the Nentsberry deposits. During the 1970s an attempt was made to recover these minerals and a small gravity treatment plant was erected near the mine entrance. Processing proved uneconomic and the venture was eventually abandoned in 1981. During this most recent phase of mining, Scraithole mine produced a number of fine specimens of sphalerite, witherite and some small examples of pale honey-coloured barytocalcite.

The Barneycraig Horse Level was a long cross-cut adit, begun by W. B. Lead in about 1760 to access veins beneath Coalcleugh, at the head of West Allendale. These veins, which form a continuation of the Dowgang-Rampgill veins of the Nenthead area, and which continue into Swinhope mine in East Allendale as the Williams vein, were associated with an extensive belt of flats in the Great Limestone. Both veins and flats carried galena-rich orebodies with an abundance of quartz, fluorite and sphalerite. Ore from Barneycraig Level was treated at a mill near the mine entrance at Carrshield. These veins were also extensively worked at higher levels from levels and shafts driven from around the tiny hamlet of Coalcleugh. The course of the Scaleburn vein may be readily followed by a line of old shaft heaps close to the road south-west of Coalcleugh. Most of these deposits were regarded as exhausted for lead ore by 1880, but the workings were reopened for zinc ore by the Vieille Montagne

Company who connected the Barneycraig workings with those of Rampgill mine at Nenthead. Zinc ore was obtained both from unworked ground left by W. B. Lead, and also from spoil back-filled by that company into old stopes and flats. Zinc ore mining continued here until 1921. Some fluorite was produced during the 1950s by reworking parts of the Coalcleugh workings.

East Allendale

Allenheads mine was the orefield's single most productive lead mine, with a recorded output of about 260,000 tons of lead concentrates to its credit. The mine, which was probably first developed in the 16th century, worked rich orebodies associated with several veins and flats which unite near Allenheads village. Early opencast and shallow shaft-workings, the remains of which may be seen both east and west of the village, were followed by deeper shafts and long cross-cut adits. The mine prospered under W. B. Lead in the 18th and 19th centuries, especially after 1806 when flats in the Great Limestone associated with Diana vein were cut. By 1822 further rich flats, in the same limestone

ALLENHEADS MINE, 'BIG DIP JUNCTION'

Group of miners and ore tubs at the 'Big Dip Junction' in Allenheads mine during the attempt at reopening, under the name of Beaumont mines, in 1979. The ore tub on the right stands at the head on an inclined drift leading to deeper parts of the mine. (Friends of Killhope Archive/ photograph by Harry Parker)

BLACKETT LEVEL

The entrance to the stone-arched Blackett Level at Allendale Town, Northumberland. This ambitious drainage tunnel, begun by W. B. Lead in 1855, was intended to pass beneath the mines at Allenheads. However, work on its driving was stopped in 1903 when it had reached 4.68 miles (7.54 km) from the entrance. (©NHMPL)

associated with the Coronation vein, were discovered. The following extract, from a series of reports by W. Crawhall, dated 1824-40, gives an impression of the richness of this ore-bearing ground:

This in all probability proved the richest length that was ever opened at Allenheads on a flat grove. I have seen this flat partially excavated several fathoms in length and 6 or 8 fathoms [11 or 15 m] in width, 14 feet [4.3 m] high, the sides and a portion of the roof appearing almost all ore, the high and middle flat were all in one, some of which would contain 0.5 cubic yards [0.4 cubic metres] with a layer of ore on the lower side 3 or 4 inches [8 or 10 cm] thick and many solid lumps of pure ore upwards of 1 hundredweight [50 kg] … cavities after cavities were broke into, full of ore ….

The Allenheads mine was largely dependent upon water power for pumping, haulage, dressing and smelting. For this, a series of reservoirs was built in the surrounding hills, with water being conducted along pipelines, down shafts and along levels. Much use was made of underground water wheels to drive pumps within the mine. A large hydraulic engine, built by the famous Tyneside engineer William Armstrong (later Lord Armstrong), and used in the mine workshops, is today preserved at the heritage centre in the village.

The development of mining at Allenheads, and indeed over much of W. B. Lead area, is inextricably linked with its most famous agent, Thomas Sopwith. So keen were W. B. Lead to secure his services that they built the impressive Allenheads Hall as his residence. It was Sopwith who, with the engineer T. J. Bewick, planned the driving of a long drainage and exploration level beneath the Allen Valley from Allendale Town, nearly seven miles to Allenheads. Starting in 1855 the Blackett Level was driven simultaneously from a number of specially sunk shafts along its course. Driving was stopped about two miles short of Allenheads in 1903. By then, although workable veins had been cut at Sipton and St Peter's, the great Allenheads mine was abandoned, having closed in 1896.

Although fluorite was reported to be abundant in the Allenheads deposits, it was never recovered during the days of lead mining. As the dumps contain surprisingly small amounts of the mineral, it was assumed, rather rashly as it turns out, that huge quantities of this mineral had probably been back-filled into the workings and left standing in unworked portions of the veins. Allenheads mine was reopened in the 1970s by the British Steel Corporation, who renamed the enterprise Beaumont mine in an attempt to rework it for fluorspar. However, despite an ambitious and very costly exploration programme, the venture proved unsuccessful and the mine closed again in 1981.

Despite its long and distinguished history and richness, Allenheads mine does not appear to have been a notable source of fine mineralogical specimens. However, some good greyish purple fluorite, some associated with black sphalerite and cream ankerite, was recovered during the 1970-81 reopening.

The fine stone mine buildings adjoining the entrance to the Fawside Level, are today in use as small business units. The stone-clad buildings erected during the 20th-century reopening of the mine remain in the centre of the village. The mine offices, adjacent to the entrance to Allenheads Hall, are currently being rebuilt as accommodation for the shooting estate. On the opposite side of the road the heavily overgrown entrance to the spiral adit, driven to give underground access to the mine ponies, can still be seen. Nearby, by the roadside, the original Gin Hill Shaft is capped by a metal grille.

North of Allenheads the remains of two smaller mines may be seen. At Sipton mine near Spartylea the shaft remains open but surrounded by a stone wall near to the stone mine buildings. At Spartylea a concrete collar surrounds the open shaft of St Peter's mine. Small, but rich, flats in the Great Limestone associated with the St Peter's vein, a vein discovered in the Blackett Level, were worked here until 1946. The mine yielded a few specimens of fluorite of an unusual beautiful bright apple-green colour. In recent years the shaft has been accessed by mine explorers who have collected many fine specimens of large brownish yellow fluorite cubes, many associated with brown ankerite.

In the tributary valley of Swinhope, to the west of Allenheads, W. B. Lead worked lead ore from the William's vein at Swinhope mine. This is an extension of one of the major veins of the Nenthead area and, like many of those veins, is distinguished by contain-

SIPTON MINE, EAST ALLENDALE, NORTHUMBERLAND

The shaft at Sipton mine, East Allendale, c.1900. It was common for W. B. Lead to enclose their shaft headframes in distinctive wooden structures of this sort. (Beamish)

ST PETER'S MINE, SPARTYLEA, EAST ALLENDALE, NORTHUMBERLAND

The headframe and ore bins at St Peter's mine, Spartylea, East Allendale, in 1930, all long demolished. This mine yielded some remarkable specimens of green fluorite and, in more recent years, has been a source of fine examples of large brownish purple fluorite cubes. (Beamish)

ing large amounts of sphalerite with little or no fluorite or barium minerals. Ambitious trials at the head of this valley during the 1950s failed to establish workable zinc deposits. Fluorite on the dumps at Swinhopehead comes from Allenheads mine spoil dumped here during the reopening work described above.

Weardale

Weardale is the name given to the valley of the River Wear upstream from the small town of Wolsingham. Included here are the tributary valleys of the Rookhope, Middlehope and Bollihope burns. Anyone with an interest in British minerals will be familiar with the names of many of the valley's numerous mines, several of which are world famous for the magnificent specimens, particularly of fluorite, produced here. As is all too often the case, however, many collections contain fine specimens simply labelled 'Weardale'.

At the very head of the dale, the Killhope mines were developed by W. B. Lead to work a group of veins, mainly in the Great Limestone. The outcrop of one of these, Old Moss vein, may be seen near the road in the banks of Killhope Burn, about 500 m

west of the entrance to Killhope Museum. The main access to the mine was through Park Level, adjacent to which the company established a large water-powered ore dressing plant. The restored ten-metre diameter water wheel, for long a gaunt rusting relic, is now one of the centrepieces of Durham County Council's Killhope Lead Mining Museum. This has been developed around the old dressing floors and offers an excellent reconstruction and interpretation of the life and work of a 19th-century northern Pennine lead mine. As restoration of parts of the underground workings to offer public access proved impossible, a new 'artificial' mine was created, accessed from the original Park Level. Designed and built with close attention to geological detail, this gives a superb impression of underground life at different periods in the area's long mining history. Above ground the museum includes a splendid collection of spectacular local minerals and houses the National Collection of spar boxes.

A short distance north of the village of Cowshill are the rather meagre remains of two of the dale's more important mines.

Burtree Pasture mine was Weardale's single largest lead producer with a total recorded output of at least

175,000 tons of lead concentrates. The major working here dates from the 19th century by W. B. Lead and their successors the Weardale Lead Company, although there is evidence for some much earlier mining. Lead mining ended at Burtree Pasture in 1890, but the mine was reopened for fluorspar mining in the 1970s which continued until economic conditions forced its closure in 1981. During this last phase of working some good specimens of purple fluorite, honey-coloured ankerite and typical 'Jack Straw' cerussite were recovered.

Immediately south of Burtree Pasture, but on the prominent E-W Sedling vein, lay the Sedling mine. Like so many Weardale mines, Sedling was originally developed as a lead mine, but became a major fluorspar producer late in the 19th century and continued as the orefield's main commercial source of this mineral until its closure in 1948. The mine buildings have long been demolished, but the shafts remain open, although fenced. The course of Sedling vein can readily be followed east of the mine site as a prominent trench extending up the fellside. From old workings near the head of this, a number of good specimens have been found of the uncommon lead manganese oxide mineral coronadite. Sedling mine produced some beautiful crystals of purple fluorite, some in very large groups.

A short distance to the west of Sedling Shafts, the Great Limestone within the structurally complex Burtreeford Disturbance has been almost completely replaced by iron minerals, oxidation of which has produced large deposits of 'limonitic' ores. Extensive old surface workings on these are striking landscape features.

The Whin Sill crops out near the river at Cowshill where it was formerly worked for roadstone at Copt Hill Quarry. Fine specimens of colourless apophyllite, collected from here in 1937 by Sir Arthur Russell, have recently been shown to be hydroxy-apophyllite. The quarry is today flooded and the faces are inaccessible.

South of the River Wear at Ireshopeburn are the remains of Weardale's last significant iron ore mine. Carricks mine, originally a lead mine developed by W. B. Lead, worked numerous veins in the Great Limestone. Associated with these were extensive siderite-rich flats from which many thousands of tonnes of iron ore were mined in the 19th century. The mine was investigated

as a source of iron ore during the Second World War and some ore was raised between 1940 and 1942.

About two miles (3.2 km) east of Carricks mine the dumps of Greenlaws mine stand above the village of Daddry Shield. Developed by W. B. Lead largely in the second half of the 19th century, the mine eventually passed to the Weardale Lead Company who worked it until 1897. Stream erosion of the large spoil heaps in recent years has revealed many large deep purple cubes of fluorite. Although strontianite has been reported from here, this seems unlikely as the specimen upon which the claim was based has been shown to be calcite. No other strontianite specimens are known from the mine.

On the north bank of the River Wear, at the hamlet of West Blackdene, fluorspar was extracted from one of the dale's best-known recently worked mines. Although there is evidence of early, perhaps medieval, working on Blackdene vein, it was during the 19th century that Blackdene mine was developed by W. B. Lead, who worked it as a lead mine until 1862. With the acquisition of Blackdene mine by the United Steel Companies in 1949, a vigorous programme of exploration and development began which resulted in this becoming one of the area's leading fluorspar producers. The workings eventually extended into the Slitt vein, the longest single vein in the orefield, and it was from oreshoots in this vein in a variety of wall-rock, including the Whin Sill, that many thousands of tonnes of fluorspar were mined. For a period in the 1970s a section of Slitt vein was encountered in which large cavities were especially numerous. The mine soon became famous for the beautifully crystallised fluorite specimens collected from these. Purple was by far the most common colour, although fine examples of green and pale yellow crystals were also found. Lustrous cubo-octahedra of galena and some good specimens of 'nail head' calcite were also collected. During the 1970s a large new treatment plant was erected at Blackdene to process both the ore from the adjoining mine and the large volume expected to come from the Beaumont mine, then being developed at Allenheads. In the event, the Beaumont mine project failed and the large new Blackdene plant was never used to its full capacity. Since the closure of the mine in the 1980s the plant has been demolished and the site cleared and landscaped.

Slitt mine, from which the vein takes its name, lies in Slitt Wood in the Middlehope Valley, north of Westgate. It was worked until 1878 by W. B. Lead. The boarded-over shaft, one of the deepest in Weardale at 178 m, lies close to Middlehope Burn. Nearby are the massive stone foundations for the winding engine and water wheel, and the remains of the bouse teams, the stone hoppers into which crude ore from the mine was tipped prior to dressing. Higher on the western side of the valley may be seen old levels, dumps and a reservoir built to power the hydraulic machinery at the mine. The course of Slitt vein may be readily traced as a prominent wide worked-out trench up the hillside on each side of the valley. Following these surface workings to the east soon leads to a series of very large opencast excavations on either side of the Westgate to Rookhope hill road.

On the west of the road are the Slitt Pasture, and on the east of the road the West Rigg opencasts. Both are large abandoned quarries for 'limonitic' iron ore, obtained here from extensive flats in the Great Limestone adjacent to Slitt vein. The West Rigg working is especially instructive. The quarry is the void left after the removal, by opencast mining, of the dark brown iron ore, a few remnants of which may still be seen on the walls of the excavation. Slitt vein here consists almost entirely of quartz with a few pockets of fluorite. It remains as a wide, unworked rib across the full length of the quarry. Narrow, slit-like stopes, from which small amounts of lead ore were extracted, may be seen in the centre of the vein. Iron ore working here dates from the second half of the 19th century. A specially built standard gauge railway took ore from West Rigg to join the inclined railway at Rookhope that connected with the Stanhope-Consett railway. An interpretation board at the roadside explains the main features of this important geological and mineralogical site which is today protected as a Site of Special Scientific Interest.

Almost two miles (3.2 km) east of West Rigg, on the north bank of the River Wear, lie the derelict remains of another of Slitt vein's major mines, Cambokeels (sometimes spelt 'Cammock Eals') mine. 'Cambo', as it was often referred to by its miners, was originally driven by W. B. Lead to de-water the Slitt vein in depth and was one of this business's many lead mines in the 19th century: the Beaumont initials may still be seen carved in the keystone above the entrance to the original horse level. The mine was reopened for fluorspar in 1906 and since then has been worked intermittently for this mineral by a variety of owners. Major development in the 1970s located substantial ore reserves that sustained the mine until its closure in the late 1980s. Cambokeels mine enjoyed the dubious distinction of being the orefield's only 'safety lamp mine', following the discovery of significant methane levels in the mine atmosphere. It was also the mine which most closely approached the Weardale granite. Indeed, not long before falling

WEST RIGG OPENCUT, WESTGATE, WEARDALE

Large ironstone flats in the Great Limestone adjacent to Slitt vein have been extracted leaving the vein, here composed mainly of quartz with a little fluorite, as a prominent ridge across the centre of the excavation. Old lead workings may be seen in the centre of the vein. (B. Young)

fluorite prices forced its closure, serious consideration was being given to following the vein down through the lowest few metres of Carboniferous rocks and into the granite. A spring of warm, mineral-rich water, cut in these deeper levels, has prompted recent investigations into the potential for commercial geothermal energy for local use.

In its later years Cambokeels mine became well-known as a source of fine specimens of fluorite in both pale purple and colourless crystals. At the deepest levels the vein locally contained large concentrations of iron sulphides, prominent amongst which was pyrrhotite, that was commonly, although very confusingly, referred to by some of the miners as 'magnetite'. Whereas much of this was in massive form, some superb bright tabular crystals up to 3 cm across, typically with curved faces, obtained here, rank amongst the best examples of this mineral ever found in Britain. Parts of the Whin Sill, cut in some of the haulage ways, contained an abundance of beautiful small colourless analcime and fluorapophyllite crystals, accompanied by a little chabazite and pectolite.

High on the north side of Weardale, above Cambokeels mine, Heights Quarry produces crushed rock aggregate and armour-stone from the Great Limestone. It has long been renowned as a source of extremely fine fluorite crystals. Purple specimens were common, but most celebrated from here were clear crystals of a deep, bottle-green colour, a substantial proportion of which contained prominent macroscopic fluid inclusions. These were found in a series of ironstone flats, similar to those at West Rigg, and which had been worked underground in the 19th century from Heights mine. They were exposed as the modern quarry cut into the abandoned workings. These old workings presented the quarry operators with considerable challenges, both of ground stability and rock quality, but also by the activities of mineral collectors. For several years a warning to collectors to keep out was prominently displayed at the quarry entrance. Quarrying has subsequently destroyed most of the remaining workings.

Numerous spectacular specimens of deep green and purple fluorite, similar in colour to those found at Heights Quarry, were recovered from a narrow vein and associated small flats in the Great Limestone in Eastgate Quarry at the former Weardale Cement Works. Like those at Heights Quarry, many of the clear crystals found here exhibited large fluid inclusions visible to the naked eye. A remarkable feature of this fluorite is its sensitivity to light. When exposed to strong sunlight, the green colour changes to dull purple within only a few weeks.

About a mile (1.6 km) up Stanhope Burn from Stanhope stands Stanhopeburn mine, the earliest records of which date from the Earl of Carlisle's workings in the 18th century. The mine was one of the few in Weardale leased by the London Lead Company who worked it until 1864. White or pale cream 'Jack Straw' crystals of cerussite were obtained from the upper, oxidised, portions of the vein. During the 20th century, under a variety of operators, the mine was an important producer of fluorspar both from unworked parts of the vein and from waste backfilled into old lead ore stopes. Nearby, small flats associated with the West Pasture vein, worked at West Pasture mine on the east side of Stanhope Burn, have yielded numerous fine specimens of fluorite with a distinctive apple-green to pale purplish-green colour. Small specimens of rosasite have also been found here.

A small vein and associated flats are exposed in the upper part of the Great Limestone at Rogerley Quarry, Frosterley. Although too small to have ever attracted interest for lead ore or fluorspar mining, the vein has for some years been worked intermittently as a source of fine mineral specimens, initially by Lindsay Greenbank and his partners, and in more recent years by a partnership of Californian-based dealers. From a short level driven into the deposit, numerous vugs up to over one metre across have yielded many magnificent specimens of clear deep bottle-green fluorite, some with a striking purplish blue 'daylight fluorescence'.

The valley of the Rookhope Burn, which joins Weardale at Eastgate, is the location of several mines well known to mineralogists the world over. Every collector of fluorite will be familiar with the name of Boltsburn mine, the site of which lies close to the Rookhope Burn on the western edge of Rookhope village. Apart from the boarded-over shaft with the foundations of its headframe and winding gear, remnants of tramways and small overgrown dumps, little now remains to recall a mine that has provided specimens to most of the world's major mineral collections. Originally developed as a lead mine by W. B. Lead from a shaft and a horse level driven into the bank of Rookhope Burn, Boltsburn became one of the last surviving lead producers of the orefield.

The mine owed its prosperity and mineralogical celebrity to the very long length of flats associated with the NE-SW trending Boltsburn vein in the Great Limestone. These flats, which in places extended for as much as 60m on either side of the vein, were discovered in 1892, and were to sustain the mine until its closure in 1932. The galena-bearing portion of the flats was generally separated from the vein by a belt of ankeritised limestone, a feature that probably explains why the main level forehead had been driven well beyond the beginning of the flat-ground before the flats were discovered. The flats consisted of ankeritised limestone in which occurred bands of galena. Large cavities, or vugs, were very common and were typically lined with beautiful fluorite cubes, in some cases up to 15cm across. Although most were purple, yellow, green and colourless specimens were also found. Many crystals were perfectly clear and transparent. The highest

quality fluorite crystals were recovered and supplied to the Zeiss Company of Austria for the manufacture of apochromatic and semiapochromatic lenses. Encrustations of ankerite and siderite are very characteristic on specimens of Boltsburn fluorite. Other minerals for which the mine is famous include quartz, galena, sphalerite, pyrite and calcite, the latter sometimes being found in stalactite-like masses composed of white 'nail head' crystals. The flats were followed along the vein for over 1.5 miles (2.4km) and were continuing strongly in the mine foreheads when increasing costs and falling lead prices forced the closure of the mine in 1932. The large dressing plant in the village, built to serve the mine, remained in operation for many years treating fluorspar from later mines in the neighbourhood. It and its associated spoil heaps have now been removed and the site landscaped.

Close to the shaft of Boltsburn mine is the site of the Rookhope Borehole, an important research borehole drilled in 1960-61 that proved the presence of the long-postulated Weardale Granite. Mineralised intersections in the core include several quartz-fluorite veins, flats in the Jew and Single Post limestones rich

BOLTSBURN MINE, ROOKHOPE, EARLY IN THE 20TH CENTURY

Boltsburn mine is one of the best known of northern Pennine mines to mineralogists for its superb crystals of purple fluorite. Beyond the mine buildings and shaft headframe, all now demolished, is Rookhope School. Still in use as the village school, this was originally built by W. B. Lead. (Beamish)

MINERS IN BOLTSBURN MINE

Miners in the famous flats in Boltsburn mine, Rookhope, Co. Durham, early in the 20th century. (Beamish)

in galena, fluorite and beautifully banded 'schalen-blende' sphalerite, a molybdenite-bearing pegmatite, and tourmaline veins in the granite. The bulk of the core is now preserved at the British Geological Survey at Keyworth, Nottingham, although small sections of it may be seen at Killhope Museum.

On the south side of Rookhope village, Stotfield (sometimes spelt 'Stotsfield') Burn mine was developed in the 19th century as a lead mine on the long WNE-ESE trending Rookhope Red vein, one of the largest veins in the orefield. It was reopened as a fluorspar producer in 1914 and remained in production until 1966, during which time it yielded many magnificent specimens of deep purple fluorite. Little remains to mark its position today. Rookhope Red vein was also worked from Redburn mine on the western outskirts of the village. This was a new mine sunk as recently as 1964 to investigate the fluorspar mining potential of a hitherto untried stretch of this major vein. Good fluorspar orebodies were worked here until 1981, since when the site has been completely cleared and returned to farmland. Some very good specimens of fluorite were found at Redburn, although the mine is also well-known for the large specimens of 'Jack Straw' cerussite obtained from the upper levels.

A little way up the valley from the site of Redburn mine stands a large stone arch, a remnant of the Rookhope, or Lintzgarth, Smelt Mill. The arch carried the flue across the road on its way to the chimney on the fell top 1.5 miles (2.4 km) to the north-west. Built and operated by W. B. Lead and its successors, the Weardale Lead Company, this was Weardale's last operating smelt mill until its closure in 1930.

The remains of several mines, together with their associated drainage levels, dressing floors and mineral tramways, may be seen along the course of the Rookhope Valley, but most conspicuous of these, near the very head of the valley, are the remains of the orefield's last major fluorspar mine.

Groverake mine is an old lead mine, the earliest workings of which probably long pre-date W. B. Lead. It was, however, this business that drove the main levels and sank the shafts during the 19th century. It stands at a point where three major veins unite near the western end of the Rookhope Red vein, but it was Groverake vein that was the first to be exploited. Although the vein was commonly up to over 10 m wide in places, records from its lead-mining days recall that it was 'chiefly fluorspar … on the whole very thinly mixed with [lead] ore'. Groverake was therefore never a very successful lead mine. It was the abundance of fluorite, in what eventually proved to be one of the largest orebodies ever found in the northern Pennines, that was to make the mine one of Britain's major 20th-century fluorspar producers under a number of operators.

Despite its size, the original Groverake orebody seems never to have been a prolific source of spectac-

REDBURN MINE, ROOKHOPE, CO. DURHAM

The shaft headframe of Redburn mine, Rookhope, Co Durham in 1981. Since its closure the mine site has been cleared and returned to farmland. (B. Young)

GROVERAKE MINE, ROOKHOPE, CO. DURHAM

In the distance, near the head of the valley, may be seen the surface plant of Frazer's Hush mine, the workings of which were connected underground with those of Groverake. (B. Young)

ular mineral specimens, although parts of the vein were noted for their beautiful banding of constituent minerals. Good quartz specimens, including some deep smoky brown crystals and some attractively banded chalcedony, were found. Studies of the Groverake ore have identified tiny crystals of bismuthinite, together with microscopic amounts of synchysite and cassiterite, giving weight to the suggestion that Groverake may lie above a deep-seated feeder centre of mineralisation. Adjacent to the vein at Groverake were extensive flats of iron ore replacing the Lower Felltop Limestone, the working of which has left conspicuous scars on both sides of the valley.

To the west of Groverake mine, a large area of 19th-century iron ore workings known as Frazer's Hushes bear witness to similar flats associated with the Greencleugh vein. During the days of lead mining, trials on this vein never seemed to have met with much success, and trials for fluorspar in the 1940s proved similarly disappointing. However, exploration in the 1970s revealed significant bodies of fluorite and led to the sinking of the Frazer's Hush and Greencleugh mines. The former working proved the most successful and was eventually connected underground with Groverake. Greencleugh was less successful and soon closed. Frazer's Hush became a noted source of fine mineral specimens. A series of small flats in the Great Limestone, cut in 1988, proved a prolific source of magnificent clear purple fluorite crystals, accompanied by some excellent bright galena crystals. The mine has also produced fine specimens of quartz, sphalerite and a few examples of acicular millerite.

A massive slump in fluorspar prices in the late 1990s rendered fluorspar mining in the northern Pennines uneconomic, and resulted in the closure of the combined Groverake-Frazer's Hush mine, sometimes referred to as Frazer's Grove mine during its final years. The Frazer's Hush site was soon landscaped, although surface buildings and a single tall steel headframe remain at Groverake.

Teesdale

The beautiful valley of the River Tees contains numerous old mines, mainly concentrated on the north side of the river and in the major tributary valleys of the Harwood and Hudeshope becks, Eggleshope Burn and the River Lune. At most of these, lead ore was the main mineral sought, although iron ore was mined and smelted at a number of places, perhaps as early as the first and second centuries AD. Commercial interest in barite and zinc ore in the dale began towards the close of the 19th century, and extraction of the former survived until late in the 20th century.

Relatively little is known about mining in Teesdale before 1752 when the London Lead Company obtained their first leases in the dale in the Hudeshope and Eggleshope valleys, north of Middleton-in-Teesdale. So successful were their operations here that they eventually established their headquarters at Middleton House, Middleton-in-Teesdale, where they also undertook major redevelopments of this small market town.

High on the hillside near the head of the Harwood Valley, and close to the Cumbrian boundary, an extensive area of old workings, collectively known as Grasshill mines, marks the outcrop of a complex of closely-spaced veins. Distinctive specimens of bright green pyromorphite, intergrown with black manganese oxides, have been found in parts of the ancient workings called Highfield Hushes.

Lady's Rake mine, by the side of the Harwood Beck, was the last mine in the dale to be worked by the London Lead Company, who abandoned it in 1902. A local syndicate of miners continued work here between 1904 and 1908, since when it has lain idle. The surface remains at 'The Lady', as the mine is known locally, include the remnants of an inclined plane which formed part of an ingenious water-balance system used for hoisting in the shaft. The remains of the pump rods and rising main still protrude from the shaft, out of which flows a steady stream of water. Parts of the dumps are scheduled as a Site of Special Scientific Interest for the presence of blocks of a unique magnetite-rich rock which locally contains patches of nickeline accompanied by galena and sphalerite. This very unusual mineral assemblage has been interpreted as evidence for mineralisation beginning here very soon after the intrusion of the Whin Sill and whilst it was still cooling.

On the hillside to the south of Lady's Rake are the remains of Willyhole and Reddycombe mines. Originally small lead mines which worked from 1852-89, the workings were reopened in 1896 and worked for zinc ore until 1912. The mines have yielded many specimens of vivid yellow greenockite as encrustations on cleavage surfaces of oxidised sphalerite.

To the south of, and parallel to, the Harwood Valley, the main valley of the River Tees flows through bleak and remote country. A substantial area of the upper valley is submerged beneath the Cow Green Reservoir, built in the late 1960s to supply water for the industries of Teesside. The reservoir conceals much of the workings of Cowgreen mine, one of Teesdale's most recently worked mines. Although the numerous veins worked here during the 19th century yielded only modest amounts of galena, all were distinguished by carrying abundant white barite, the mineral that was to sustain mining here until 1952. Barite mining began in 1902 and continued until 1919 under the Hedworth Barium Company. In 1935 the Wrentnall Baryta Company, then operating in Shropshire and Ayrshire, acquired the mine. They sank a new shaft, called Wrentnall Shaft, and built a row of miners'

cottages, which they named Wrentnall Cottages, at Langdon Beck. Barite was extracted both from opencast and underground workings. The veins at Cowgreen cut some of the lowest beds of the local Carboniferous succession, including the thick Melmerby Scar Limestone. This has here been converted by contact metamorphism adjacent to the Whin Sill into a coarse-grained marble, known from its distinctive crumbly weathering as 'sugar limestone'. The thin soils developed on this rock at Cowgreen support an extremely important relic arctic alpine flora, including the beautiful and rare spring gentian (*Gentiana verna*), for which the site is famous. Cowgreen, and much of Upper Teesdale, is designated as a National Nature Reserve for its botanical and geological interest. Cowgreen mine never seems to have been a major source of fine mineral specimens, although some good clear tabular barite crystals were recovered and Sir Arthur Russell obtained small crystals of millerite from the mine in 1939. Quartz occurs locally as white sugary aggregates, thought to be replacements of marble. Although much of the mine site lies beneath the reservoir, a few opencast workings and some level entrances may be seen near the modern car park.

Remotely situated on Cronkley Fell, south of Cowgreen, are several old mines including one of the

MINERS IN STOPE, COWGREEN MINE, CO. DURHAM, TEESDALE

Miners drilling in a stope in one of the barite veins at Cowgreen mine, Teesdale in 1949. (Friends of Killhope Archive, Killhope Lead Mining Museum, Weardale)

orefield's two Silverband mines, Yorkshire Silverband mine. Lead mining is known to have taken place here in the 19th century, although much earlier working is likely. A fine set of primitive ore-crushing rollers with rough stone flywheels, which stood for a long time at the mine, are now at Killhope Lead Mining Museum in Weardale. Good specimens of white fibrous aragonite have been found here, as well as a few examples of grass-green pyromorphite.

The highlight of Teesdale for most visitors is High Force, an impressive waterfall where the River Tees plunges over a ledge formed by the Whin Sill. Concealed from view a few hundred metres upstream is Force Garth Quarry (sometimes called Middleton Quarry), from which Whin Sill dolerite is extracted for roadstone. From time to time excellent specimens of white or pale pink pectolite are found in joints in the sill, and some examples of pale green prehnite and a few crystals of chabazite have also been found here. On the north side of the road immediately west of High Force stands the long-abandoned High Force Quarry which was the source of fine pectolite specimens seen in many collections. Analcime, chabazite and stilbite have also been found here. High Force Quarry is also noted for dolerite pegmatite veins found near the centre of the sill. These striking rocks, which are also found at Force Garth, contain bladed crystals of clinopyroxene up to several centimetres long.

In the small tributary valley of Ettersgill, attempts have been made to mine zinc ore from small flats in the Single Post Limestone at the curiously named Dirt Pit mine. Although the most recent trials, made during the Second World War, identified an orebody from which a high grade zinc concentrate could be produced, the deposit was never mined. Several small veins and flats which, like those at Dirt Pit, are characterised by an abundance of siderite and sphalerite and the absence of galena, fluorite and barium minerals, occur on the south bank of the River Tees near Wynch Bridge. Several trial workings, visible from the riverside footpath, were made here, but no mining followed.

Pike Law mines comprise a striking group of large hushes and associated shaft workings on either side of the Newbiggin to Westgate road, in the headwaters of Bowlees Beck, north of Bowlees. Mineralisation, in several veins and associated flats in the Great Lime-

PIKE LAW MINES, TEESDALE, CO. DURHAM
A large complex of deep opencast workings and hushes scar the heavily mineralised outcrop of the Great Limestone at Pike Law mines, Newbiggin Common, Teesdale. (©NHMPL)

stone, is well exposed in places here. Good crystals of purple to grey fluorite have been found, although the site is perhaps best known for comparatively large crystals of anglesite up to 15 mm long, collected from oxidised portions of the orebodies. The spoil locally contains clay ironstone septarian nodules, the septa of which are filled with quartz. Removal of the ironstone by weathering has produced some very curious honeycomb – like masses of quartz, known rather inexplicably as 'beetle stones'. It should be noted that in recent years specimens of fluorite and anglesite collected here have been passed into collections incorrectly labelled 'Flushiemere Mine'. The veins worked at Flushiemere mine, which is on the eastern side of the valley of Bowlees Beck, lie in the barium zone of the orefield and were worked as recently as 1946 for barites. A number of fairly good specimens of pyramidal crystals of witherite have been obtained from this mine.

The profile of Hardberry Hill, north-east of Newbiggin, is distinguished in most views by the conspicuous notch of Coldberry Gutter. This huge opencast working, one of the largest in the orefield, was excavated along the western extremity of the Lodgesike-

COLDBERRY MINE, NEAR MIDDLETON-IN-TEESDALE CO. DURHAM, 1932

The waterwheel drove the crushers; a water turbine and small steam engine powered the jigs and buddles in the building. (Beamish)

Manorgill vein, part of a major vein complex on the northern side of Teesdale. Often referred to as a 'hush', the gutter crosses the watershed between the valleys of the Bowlees and Hudeshope Becks. Whereas 'hushing' clearly played some part in its excavation, as witnessed by the remains of reservoirs on the fellside to the north, it is most likely that much of the enormous volume of rock excavated here was removed by conventional opencast mining. The date of the surface working is unknown, although the London Lead Company worked the veins underground from Red Grooves and Coldberry mines in the second half of the 19th century. Exploration of the mines in the 20th century yielded a little lead ore, the last such working ending in 1955. Clay ironstone nodules from some of the shale beds exposed in Coldberry Gutter have yielded specimens of millerite similar to those found in the South Wales coalfield. Further north-east the London Lead Company worked the large Lodgesike, Wiregill and California mines on the outcrop of the same vein complex until 1902, after which the Wiregill Deep Mining Company continued to operate parts of the mines until 1913. Subsequent attempts to recover fluorspar from the spoil heaps from these mines, and from the workings at Coldberry Gutter, all proved unsuccessful. Old hushes, large spoil heaps and mine reservoirs are all that remain to mark these once important centres of lead mining.

Teesdale is joined at Mickleton by Lunedale, near the head of which are several old mines. Lunehead mine was worked by the London Lead Company between 1770 and 1880, although its origins may be much older. It was never a rich lead mine, but between 1884 and 1937 it was an important barite producer under the Reynoldson family. Further attempts at barite production were made in the 1960s by Athole G. Allen Ltd, who were then working the nearby Closehouse mine. Parts of the Lunehead workings have yielded fine specimens of tabular barite crystals. Specimens of beautiful pale sky-blue and white banded aragonite, and more recently numerous excellent specimens of colourless barytocalcite crystals, have been collected from the Forster's Hush workings at the extreme east end of the mine.

About 1.5 miles (2.4 km) north of Lunehead mine lies Closehouse mine, Teesdale's most recently worked barite mine. The deposits here lie within the Lunedale fault-system, the major structural feature that forms the southern boundary of the Alston Block. The fault-system has here been intruded by a wide dolerite dyke that forms part of the Whin Sill suite of intrusions. Complex alteration of the dolerite, and its subsequent replacement by barite, has formed large bodies of that mineral. The deposits at Closehouse attracted exploration at an early date and huge hushes were already in existence when the London Lead Company acquired the lease in 1770. Their exploration, including the driving of levels beneath the hushes, showed only very poor lead values but revealed huge quantities of barite. Athole G. Allen Ltd obtained the lease in 1939 and began underground barite mining in 1945 in a wide body of barite, in part replacing dolerite. So wide was the orebody that in places room-and-pillar mining was employed. Since the 1970s, under successive owners, opencast mining was employed to extract barite, leaving a spectacular opencast excavation on the western side of the Arngill Valley. Working eventually became uneconomic and extraction ended in 2001. From time to time Closehouse has produced interesting mineral specimens, including some fine water-clear barite crystals. Good small grass-green crystals of pyromorphite have long been well-known from here, and more recently the site has yielded fine specimens of rosasite, aurichalcite, anglesite and a few small examples of leadhillite, the first to be confirmed from a Pennine locality.

The Derwent Valley

In the headwaters of the River Derwent a number of old lead mines lie clustered around, and to the south of, the tiny villages of Blanchland and Hunstanworth. Although Blanchland can trace its origins back to the 12th-century monastery, the outline of which is retained in the present village, the earliest definite record of mining nearby dates from 1475 when Shildon mines, immediately north of the village, are known to have been active. These mines enjoyed a long history extending into the early 19th century, when a Boulton and Watt pumping engine, the stone-built engine house of which may still be seen, was installed. An interesting feature of these mines is the abundance in the veins of beautifully banded pale blue, yellow and white chalcedony, some of which pseudomorphs earlier fluorite.

The largest and most productive mines of this area lay along a group of veins south of Hunstanworth. Many of these were held by the London Lead Company during the 18th century, followed by a number

of other concerns in the 19th century. Old dumps, water-courses and a well-preserved smelt mill flue are still conspicuous in the Boltsburn Valley (not to be confused with the Boltsburn of the Rookhope area further south). Although the veins here were commonly wide, they typically carried rather poor lead values. The galena was, however, commonly richer in silver (typically around 500 ppm [17.5 oz of silver per ton of lead]) than many other parts of the orefield. In the 1920s mining was resumed for fluorspar and, under a succession of operators, mining continued here until the closure of Whiteheaps mine in the late 1980s.

The Tyne Valley

A small number of significant mines within the Tyne Valley, south of Hadrian's Wall, between Haydon Bridge and Hexham, worked veins which are generally regarded as comprising an outlying part of the main northern Pennine orefield. All lie within the outer, barium zone, of the orefield.

Close to Hadrian's Wall lay Settlingstones mine, one of the orefield's most famous mines. The roughly NE-SW trending Settlingstones vein was known in the latter part of the 17th century and worked as a lead mine until 1873. At that date, mining advanced southwestwards through a cross-vein beyond which the vein filling suddenly changed to witherite with only very

CHALCEDONY, SHILDON MINES

Banded white and pale blue chalcedony, typical of the form of the mineral that was so abundant in some of the veins in this part of the Derwent Valley. This specimen is 7 cm across. (©NERC)

'PICKING BELT', SETTLINGSTONES MINE, NEWBROUGH, NORTHUMBERLAND, c.1930

Men and boys sorting ore on the 'picking belt' at Settlingstones mine. (Beamish)

MINERS AT WINTER'S SHAFT, SETTLINGSTONES MINE, NEWBROUGH, NORTHUMBERLAND, 1905

Note the bunches of candles and loops of blasting fuse carried by some of the group. (Beamish)

small amounts of galena. Production rapidly changed to witherite, which was to bring international fame to the mine and to sustain it until its closure in 1969. For long periods Settlingstones was the world's sole commercial source of this unusual mineral. The ore-body was remarkable for its width, up to 9 m in places, and the purity of its witherite filling. From it were recovered many beautiful specimens of pale cream complex twinned crystals of witherite, some overgrown with crusts of tiny cruciform crystals of harmotome. Pale lime-green strontianite was found sparingly. Large cavities lined with groups of very distinctive chisel-shaped barite crystals were extremely common. A rich pocket of nickeline, associated with a little ullmanite, was found in the 1920s, fine specimens of which were recovered by Sir Arthur Russell. Apart from a pipe ventilating the sealed Frederick Shaft by the roadside, the row of miners' cottages and the increasingly over-grown dumps (today scheduled as a Site of Special Scientific Interest for their mineral content) by the Settlingstones Burn, little now remains to mark this once important mine.

About one mile (1.6 km) east of Settlingstones, and on a sub-parallel vein system, stood the Stonecroft and Greyside mines, the former site conspicuous today by the massive red brick walls of the Boulton and Watt

engine house. The veins here were discovered only in about 1853 and were worked for lead until 1896. Sphalerite, which was abundant here, was never re-covered during lead mining; it was extracted from the spoil heaps during the early 1950s. Unusually for the northern Pennines, much of this mineral here is very pale brown in colour and is commonly found in a coarsely banded 'schalenblende' form. Good small cruciform harmotome crystals were found here by Sir Arthur Russell.

Similar pale brown banded sphalerite is common at Langley Barony mine, another large lead mine that enjoyed a rich but brief career during the late 19th century, about two miles (3.2 km) south-west of Settlingstones.

North of Hexham, on the outskirts of the village of Acomb, are the relatively meagre remains of the Fallowfield mine. Like Settlingstones, this mine worked a roughly NE-SW trending vein. The mine is known to have been in existence in 1611 and was worked solely for lead ore up to 1846. At this time production changed to witherite, which was abundant in the vein. From this mine have come many magnificent crystal-lised specimens of this mineral. Probably most com-mon in old collections are pseudo-hexagonal pyramids or bipyramids, although stout prismatic crystals and mammillated masses were also common. Fallowfield was one of the sources of the original specimens of alstonite, first collected here in the 1830s. White, colourless, or more rarely, pale rose-pink, bipyramidal crystals of this mineral are commonly associated with witherite crystals. Much of the mine site today lies beneath a caravan park, although a few small buildings remain, notably the old engine house and chimney, now a private house, adjoining the New Engine Shaft. The few remaining spoil heaps are scheduled as a Site of Special Scientific Interest.

South of the River Tyne, in a woodland setting near Langley Castle, lies the hamlet of Langley, the site of one of the area's largest and most important smelt mills. Built in 1767 by the Commissioners of Greenwich Hospital, to smelt both lead and zinc ores from the mines of Alston Moor, in its early days Langley Smelt Mill was said to be one of the largest of its type in Europe. A 19th-century writer described it as resembling a small village. Like most northern Pennine smelt mills, Langley featured a long flue up the hillside to the chimney on the hill top which remains such a prominent local landmark. These flues were built to condense and recover lead from the smelt mill fumes. Since the mill's demolition in 1887 the site has become a peaceful woodland, although a few buildings remain. The course of the flues, including the bridge that carried it over the Hexham and Allendale railway, may be seen in the grounds of the Garden Station teashop and gardens.

The Northumberland and Durham Coalfield

A few of the numerous faults within the coalfield that lies to the east of the northern Pennines carry an abundance of barium mineralisation. Barite and witherite were worked alongside coal as commercial products at South Moor, Ushaw Moor and New Brancepeth collieries in County Durham. Fine examples of witherite were obtained from all of these with, in addition, a few specimens of alstonite at New Brancepeth colliery.

A curious feature at a handful of collieries in southern Northumberland was the abundance of barium chloride in the water pumped from the mines. Such waters have been considered to represent residual mineralising brines, perhaps related to the northern

'SUNDAY STONE'

Fine-grained barite with intermittent bands rich in coal dust, formed inside a pumping pipe. (©NHMPL)

Pennine mineralisation. Reaction of these fluids with sulphate-rich groundwater caused precipitation of barite which built up as layers coating the inside of the square section wooden pump pipes in the mines. Known to the miners as 'Sunday stones', examples of these deposits are present in several collections of northern England minerals. They consist of bands of fine-grained white barite alternating with layers of black coal dust. Typically six narrow white layers correspond with six week nights, and six narrow black bands with the working days when coal dust was settling in the mine water. A thick white layer was deposited on Sundays when there was no coal production. Weeks with holidays were indicated by fewer black bands.

So abundant were these barium chloride-rich mine waters that at Eccles colliery, north of Newcastle-upon-Tyne, barium sulphate, known as 'blanc fixe', was produced commercially from the mine waters between 1937 and 1969.

The Northern Pennine Orefield – The Askrigg Block

As in the Alston Block, the date of the earliest mining is unknown. However, as in that area, there is some evidence that the Brigantes, the tribe that inhabited the area at the time of the Roman invasion, may have worked metals. Evidence of Roman working of lead in the Yorkshire dales is provided by the discovery in the Greenhow and Swaledale areas of pigs of lead stamped with Roman marks. Whereas it is popularly supposed that the Jackass and Sam Oon Levels at Greenhow may be of Roman origin, no workings of undoubted Roman date have yet been identified. It is known that St Wilfred used sheets of lead on the roof of York Minster, suggesting that some lead may have been mined and smelted in Yorkshire during the 7th century. By the Norman period, lead and iron mining is recorded in monastic records for Wensleydale and parts of Swaledale, and thereafter documentary references to mining within the Yorkshire Pennines become more common.

A major change in ownership of mineral rights took place at the dissolution of the monasteries under Henry VIII, when lands hitherto controlled by the great abbeys at Fountains and Bolton were confiscated and passed eventually into private hands. Mines in the Grassington and Wharfedale areas became the property of Lord Burlington, whose descendants, the Dukes of Devonshire, still hold them. This family, who were also owners of very substantial mineral rights in Derbyshire and Devonshire, undertook major developments of these properties during the 17th century. Miners who were brought from the Derbyshire mines at this time introduced a number of Derbyshire customs and practices into these Yorkshire mines.

In Wensleydale, mining on Lord Bolton's land was active at this time. Further north, in Swaledale, Lord Wharton promoted active mining and exploration on his estates by small partnerships. A reorganisation of mining on Lord Wharton's lands took place in the 1670s. At this time, a number of the small partnerships were dispossessed in order to encourage mining on a larger scale, perhaps influenced by Hochstetter's methods in the Lake District.

Like the Earl of Derwentwater in the Alston Moor area, Lord Wharton was deprived of his estates as punishment for his part in the rebellion of 1715. His estates were sold by the Crown and in 1760 became the property of Anne Draycott. It is believed that it was her initials that came to be used by the 'A. D. Mining Company', formed in 1873, although the mines were referred to as the A. D. mines long before the establishment of this company.

The Arkengarthdale Royalty, which was bought in 1656 by Dr J. Bathurst, physician to Oliver Cromwell, passed to his grandson Charles Bathurst who gave his

1	High Longrigg mine	44	Swaledale Providence mine
2	Hartley Birkett mine	45	Spout Gill Hush
3	Stennerskeugh Clouds mines	46	Beezy mine
4	Cumpston Hill mines	47	Friar's Intake mine
5	Great Sleddale Copper mine	48	Summer Lodge mines
6	Lane End mine	49	Stags Fell mine
7	Little Moor Foot mine	50	Sargill mine
8	Keldside mine	51	Raygill mine
9	Beldi Hill mines & smelt mill	52	Warton Hall Level
10	East Arn Gill mine, including Adelaide Level	53	Wet Groves mine
		54	Virgin Moss mine
11	Swinnergill mine	55	Woodale mine
12	Lownathwaite mine	56	Whitaside mine
13	Blakethwaite mine & smelt mill	57	Apedale mine
		58	Harkerside mine
14	Friarfold Hush	59	Grinton mine
15	Old Rake Hush	60	Grinton Smelt Mill
16	Bunton Level	61	Ellerton Moor mine
17	Sir Francis Level	62	Cobscar Smelt Mill
18	Brandy Bottle mine	63	Keld Heads mine
19	North Rake Hush	64	West Burton mine
20	Old Gang mines & smelt mill	65	Thoralby mines
		66	Bishopdale Gavel mine
21	Barras End mine	67	Buckden Gavel mine
22	Surrender Smelt Mill	68	Starbotton Moor End mine
23	Surrender mines	69	Kettlewell Smelt Mill
24	Turf Moor Hush	70	Old Providence mine
25	Hungry Hushes	71	Conistone Moor mine
26	Octagon Smelt Mill	72	High Mark mines
27	Langthwaite Smelt Mill	73	Malham & Pikedaw mines
28	Faggergill mine	74	Meal Bank Quarry
29	Stang mine	75	Grassington Moor mines & smelt mill
30	Windegg mines		
31	Moulds mine	76	Yarnbury mines
32	Fell End mines	77	Lolly mine
33	Hurst mines	78	Merryfield mine
34	Copperthwaite mine	79	Prosperous-Providence mine
35	Feldom mines	80	Cockhill mine & smelt mill
36	Whashton Smelt Mill	81	Jackass Level
37	Sorrowful Hill mine	82	Coldstones Quarry
38	Forcett Quarry	83	Duck Street (Greenhow) Quarry
39	Merrybent mine		
40	Middleton Tyas mines	84	Greenhow Hill mines
41	Gingerfield mine	85	Blackhill mine
42	Billy Bank mine	86	Burhill mine
43	Lover Gill mine	87	Trollers Gill mine
		88	Appletreewick mine

The Askrigg Block

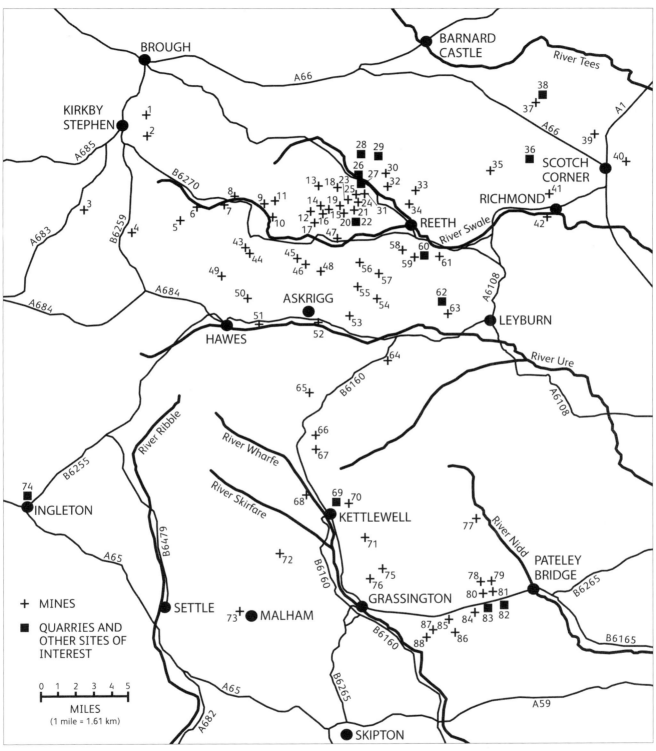

'BELL PITS' AROUND YARNBURY,
NEAR GRASSINGTON, NORTH YORKSHIRE

The characteristic shape of old shafts, sometimes referred to as 'bell pits', surrounded by low circular mounds of spoil, mark the course of veins on the ground around Yarnbury. (©NHMPL)

JACKASS LEVEL, GREENHOW HILL, NORTH YORKSHIRE

The entrance to this narrow 'coffin level' is cut into hard gritstone. A clear 'herring bone' pattern of pick marks on the walls of the level betray its hand-cut origins in pre-gunpowder days, possible as early as the 16th century. Traces of these characteristic pick marks can be seen in this picture, immediately above the entrance. (©NHMPL)

initials to the 'C. B. Mines'. This company came to dominate lead mining in this part of the orefield until the early years of the 20th century.

The London Lead Company, one of the major influences in mining in the Alston Block, acquired leases of mines on the south side of Swaledale in 1733 and in parts of Wensleydale in 1734, although their operations in this part of the Pennines were modest compared to their activities further north.

Although lead ore was the main mineral originally mined from this area, small centres of copper mineralisation were worked on a small scale as early as the 15th century, notably near Kirkby Stephen, Malham and Richmond. Deposits at the latter location may be regarded as part of a remarkable concentration of small, but extremely rich, deposits of copper ore worked for about 30 years in the 18th century around the village of Middleton Tyas on the eastern fringe of the orefield.

Although widely present in the Askrigg Block, zinc ores never achieved the importance they assumed in the Alston Block. However, a unique deposit of smithsonite, known locally as 'calamine', was worked in the latter years of the 18th and first few years of the 19th century at Malham.

As in other British mining fields, the dramatic fall in lead prices during the last 20 years of the 19th century proved disastrous for the area's mines. Mining for fluorspar, barite, and locally a little witherite, was undertaken during the 20th century at a few mines, but it has never been on the scale, or achieved the same success, as that seen in the Alston Block. Small-scale reprocessing of old spoil heaps in Swaledale to recover barite, which ended in the 1980s, was the last commercial working of vein minerals in the Askrigg Block. Despite the 19th-century collapse of large-scale mining, several parts of the Askrigg Block have, during the 20th century, attracted exploration for further metal deposits. Investigations, including exploratory drilling, in the Gunnerside area of Swaledale, at Middleton Tyas and in the area of the Craven Fault, failed to reveal new workable deposits.

The long history of lead mining in the Askrigg Block has left a considerable legacy of remains. The most extensive and spectacular hushes in the Pennines are to be seen in parts of Swaledale. Spoil heaps, some very large, mark the sites of major mines and extensive spreads of spoil, and tailings from early dressing activities cover huge areas of fellside along the outcrops of several veins. The remains of smelt mills, their flues, and other associated structures, are well-known landscape features, particularly in Swaledale.

Swaledale

The beautiful North Yorkshire valley of the River Swale includes some of the most intensely mineralised ground in the Askrigg Block. A complex elongate belt of closely-spaced mineral veins, collectively termed the 'North Swaledale Mineral Belt', extends across the hills of the north side of the valley from near its head eastwards to near Richmond. It crosses the lower reaches of the tributary valley of Arkengarthdale at Langthwaite. A much smaller concentration of veins occurs on the south side of the valley, many crossing the watershed into Wensleydale to the south.

In a remote moorland setting, high in the tributary valley of Great Sleddale, is a small trial working of unknown date. Some good amber fluorite crystals are present, but the only obvious metallic minerals are azurite and malachite, some in very good, but small, crystals, together with traces of chalcopyrite. Near the junction of Great Sleddale Beck with the Swale are the meagre remains of Lane End mine, a small lead mine known to have been active in the 18th and 19th centuries, at which galena was accompanied by witherite, with traces of barytocalcite and strontianite.

About one mile east of the tiny village of Keld, the steep northern slopes of the dale are scarred by the workings of Beldi Hill mines. Remnants visible today include old hushes and spoil heaps, together with ruined buildings and the site of a smelt mill. Mining was in progress here in 1738, and it is understood that some reworking took place as recently as 1846. The dumps, especially those from the Crackpot Level, are of interest for the local abundance of small sprays of white barytocalcite crystals. South of Beldi Hill, the Adelaide Level at East Arn Gill mine at one time yielded good greenish yellow cubes of fluorite from a small flat within the Underset Limestone.

The deep valley of Gunnerside Gill that joins the Swale at Gunnerside, cuts numerous roughly E-W veins.

OLD GANG SMELT MILL, HARD LEVEL GILL, SWALEDALE, NORTH YORKSHIRE

Beyond the mill ruins may be seen the remains of the flue leading up the fellside to a chimney on the fell top. (©NHMPL)

Several of these have been worked opencast in a series of spectacular hushes on both sides of the valley about 1.5 miles (2.4 km) north of the village. Levels have been driven into the veins from these hushes. The huge piles of spoil yield interesting mineral specimens. Primary minerals include barite, witherite, fluorite and a little strontianite. Here, as at many mines in Swaledale, small bright red earthy patches of cinnabar occur in blocks of oxidised veinstone. These veins were intersected at depth in the Sir Francis Level. Driving of this level, which was named after Sir Francis Denys, son of the proprietor, began in 1864. This ambitious project was intended to de-water the mines then being worked separately by the A. D. and Old Gang companies, who shared the cost of driving. The work proved difficult and expensive because of the hardness of the rock. Compressed-air rock drills were eventually employed: the air receiver for the drills still stands at the level entrance. From within the level, a shaft was sunk to explore deeper parts of the veins. The large hydraulic pumping engine installed here remains *in situ* and is much-visited by mine explorers. Large dressing floors were set up close to the level mouth to treat the large volumes of crude ore brought to the surface. Late in the 19th century, dressed ore from the Sir Francis Level

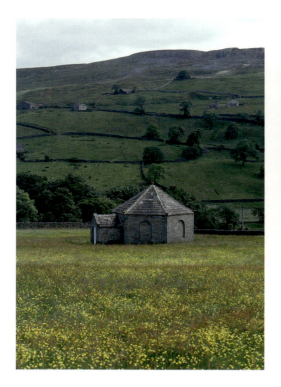

LEFT: C. B. MINES POWDER STORE, LANGTHWAITE, NORTH YORKSHIRE

The well-preserved hexagonal C. B. Mines powder store, built about 1804, at Langthwaite in Arkengarthdale, North Yorkshire. (©NHMPL)

ABOVE: FRIARFOLD RAKE, HARD LEVEL GILL, SWALEDALE, NORTH YORKSHIRE

Extensive spoil heaps mark the course of Friarfold Rake and associated veins on the high moorland at the head of Hard Level Gill, Swaledale, North Yorkshire. (B. Young)

floors was taken back underground in the Bunting Level on the east side of the valley and trammed approximately six miles (9.6 km) underground to emerge at the Hard Level, adjacent to the Old Gang Smelt Mill. This rather unusual procedure avoided the more costly surface transport of ore across the fells. The Sir Francis Level workings have yielded striking examples of radiating white strontianite crystals.

Friarfold, North Rake and Old Rake veins may be easily followed eastwards from Gunnerside Gill through the extensive spreads of surface spoil and almost continuous lines of hushes and old shafts along their courses. Until the 1980s parts of the spoil heaps from the Old Rake workings were being reworked for their barite content at a small treatment plant near Old Gang Smelt Mill. A feature of these dumps, and indeed most of the dumps in this part of Swaledale, is the abundance within them of barite in the form of rather cellular masses of small chisel-shaped crystals. This is a form generally regarded as indicative of formation from an earlier barium carbonate mineral. Witherite is also common locally. Dull green pyromorphite, an unusual mineral in the Yorkshire Pennines, occurs sparingly on the Old Rake dumps, and small specimens of cinnabar and rosasite have also been found.

At the head of Hard Level Gill occurs a bewildering array of old spoil heaps from the many workings on the cluster of veins that cross the gill here. Sites of particular interest include the Brandy Bottle mine, which was worked from two steeply inclined adits driven soon after 1814. Fine examples of delicate plate-like cerussite crystals were recovered from here several years ago. The courses of the prominent North Rake and Old Rake veins are marked by large hushes on the west side of the valley.

Further down Hard Level Gill are the gaunt but striking remains of the Old Gang Smelt Mill. Lead smelting is known to have been carried out here as early as 1771, although the building seen today was built after 1828 and worked until about 1890. Notable features of this large smelt mill include the remains of the peat house, in which a year's supply of peat fuel was stored, on the hillside above the mill, and the flue which runs to a chimney about a half mile (0.8 km) further up the hill. On the opposite bank of the stream from the mill is the huge dump from one of the area's most extensive sets of underground workings. These were accessed via the Hard Level, driving of which by Lord Pomfret began in 1785. Good examples of the district's most important minerals, including galena, sphalerite, barite, witherite and strontianite, occur on this and the nearby dumps. A little further down the valley, in the stream known as Ashpot Gutter, are the conspicuous dumps from the Victoria Level, driven to several important veins in 1859. Specimens of a mixture of witherite and strontianite, known locally as 'water spar', are said to have been collected here, although none is visible today. Nearby, at the Barras End mine, striking specimens of brown banded barite, reminiscent of the well-known 'Derbyshire Oakstone', have recently been found.

A little to the east of Surrender Bridge, where the Low Row to Langthwaite road crosses Hard Level Gill, are the ruins of another of the dale's famous lead smelting mills. Surrender Mill, which smelted ore from many of the A. D. Company's mines, was built in 1839 as a replacement of earlier mills on the site, which date back to the late 17th century. The slag heaps at Surrender have yielded an interesting suite of so-called 'slag minerals' formed by the weathering of the metal-rich slag.

Almost a mile (1.6 km) further north, near Bleaberry Gill, stood the Barras End group of mines, old lead mines that were reopened briefly in about 1925 in an attempt to mine for witherite. Beautiful pseudo-hexagonal crystals of this mineral from the nearby Danby Level are in the Lady Anne Cox Hippesley collection, now part of the Russell collection, in the Natural History Museum.

The hillside of Great Pinseat, between Bleaberry Gill and Arkengarthdale, is spectacularly scarred by a vast array of hushes, collectively known as Hungry Hushes, excavated on numerous strong veins of the North Swaledale mineral belt. It has been said that the hushes on the Stodart and Blackside veins comprise some of the wildest and most impressive mining landscapes in the northern Pennines. Important amounts of lead ore were obtained here, although the name 'Hungry Hushes' suggests that some at least may have proved disappointing. A short distance further south, and immediately west of the road, are several other major hushes, most notably the huge Turf Moor hush. Extensive underground mining from levels also took place here under the C. B. Company. The main underground workings were accessed from the Moulds Old Level, the huge spoil heap which is crossed by the Low

TURF MOOR HUSH, ARKENGARTHDALE
One of the many substantial opencast workings on the veins of this area. (©NHMPL)

Row to Langthwaite road. The lead-mining landscape here is to some extent modified by the effects of 20th-century chert mining on the same hillside.

To the west of the road, just south of Langthwaite village, are the remains of Langthwaite Smelt Mill, built in 1824. The double flue up the hillside to the west combines with the old flue from the old Octagon mill which formerly stood on the opposite side of the road to the Langthwaite mill. Nearby, in the middle of a grass field, is the hexagonal powder house of the C. B. Company.

North of Langthwaite lie the Faggergill and Stang mines. Mining here may date from before the 19th century, but the main period of working was between 1840 and 1881. These were the last substantial mines in Swaledale to work, with final abandonment coming in 1913. Notable specimens of 'flos ferri' aragonite are known from the Faggergill workings.

East of Langthwaite the veins of the North Swaledale mineral belt may be traced through the large hushes and underground workings of the Fell End mines, across Hurst Moor to the complex of old workings of the Hurst mines at the head of Shaw Beck. Dumps from workings on Copperthwaite vein on Marrick

Moor contain an abundance of hemimorphite, commonly in well-crystallised masses.

Veins, and therefore mines, are much less numerous on the southern side of Swaledale.

Old workings on the Lover Gill and Swaledale Providence veins may be seen to the east of Butter Tubs Pass, south of Thwaite. A complex of veins was worked during the 18th and 19th centuries at the Spout Gill and Summer Lodge mines on either side of Oxnop Gill. Some of the ores worked here were very rich and were smelted in a mill in Summer Lodge Gill. Records of mining on several of the veins south of Gunnerside and Grinton at Whitaside, Harkerside and Grovebeck, extend back until at least the 16th century, and during subsequent years they are known to have been the subject of several disputes. Some of the workings were leased by the London Lead Company for a comparatively short period during the 18th century, one of this company's few incursions into this part of the Pennines. Grinton Smelt Mill and its associated flue, in Cogden Gill about a mile (1.6 km) south of Grinton village, is perhaps the best preserved smelt mill building remaining in the Yorkshire Pennines.

Old workings for copper and lead ore, probably dating from the 17th and 18th centuries, mark the outcrop of Feldom vein north-west of Richmond; but as this upland is part of a military training area, access is restricted.

Wensleydale

The old name for the valley of the River Ure, 'Yoredale', today known as Wensleydale, is preserved in geological literature as the name for the characteristic Carboniferous repeated cyclical sequence of limestones, mudstones and sandstones, known as the Yoredale facies. The distinctly terraced hillsides formed by differential erosion of these rock units are characteristic features of this beautiful dale.

Compared to Swaledale, Wensleydale carries comparatively few metalliferous veins, though there were some formerly important mines, the names of several of which are familiar to mineralogists. A number of Wensleydale's mines, including the Arn Gill, Raygill, Thornton and Warton mines, worked veins which cross the watershed from Swaledale.

About a mile (1.6 km) north of the picturesque village of Aysgarth, below the crag known as Ivy Scar, are the workings of Wet Groves mine. This relatively small mine worked lead ore from a series of veins and flats, both opencast and underground. The mines are known to have been leased in 1754, but some of the workings were then already very old. They were worked again during the 19th century until their final abandonment in 1870. The Wet Groves workings have become famous for magnificent groups of sharp, chisel-shaped white, cream or colourless barite crystals, collected mainly in the 1970s and 80s. Good examples of fluorite, witherite, hemimorphite and aurichalcite have also been found here.

Remotely situated on Bolton Moor, north-west of Castle Bolton, are the remains of Virgin Moss mine, the earliest workings of which are probably very old. Mining is, however, known to have been in progress here in the 19th century when, in addition to lead ore, some barite and a little witherite are understood to have been sold. A flat containing witherite was present in the workings, and a pile of picked lumps of very pure witherite remained for many years on the old sorting floors. Excellent specimens of fibrous crystalline pale yellow witherite have been collected here. The mine is also of interest for the presence of the rare zinc-rich kaolinite-serpentinite-group mineral

GOETHITE PSEUDOMORPHS, VIRGIN MOSS MINE, WENSLEYDALE, NORTH YORKSHIRE

Goethite pseudomorphs after crystals of marcasite from Virgin Moss mine. Although iron sulphide minerals are not abundant in the Askrigg Pennines, excellent pseudomorphs after marcasite have been found at several mines in Swaledale and Wensleydale. (B. Young)

fraipontite, which occurs here as pale green to brown banded masses intergrown with gibbsite. Good specimens of hemimorphite and fine goethite pseudomorphs after spear-shaped marcasite crystals have also been found.

North-east of Castle Bolton a series of old workings clearly marks the outcrop of the vein known as Cobscar Rake. Good specimens of hemimorphite have been collected from the dumps. Cobscar Smelt Mill was built near the eastern end of the workings in 1765, but is now almost totally destroyed. Near the site of this mill are the remains of the flues from the Keld Heads Smelt Mill which was sited further down the valley sides near Keld Heads mine which supplied it with ore. This mine, the largest and most important in Wensleydale, and said to have been the richest in Yorkshire, worked principally in the vein called Chaytor Rake, a strong NNW-SSE trending vein which connects with some of the south Swaledale veins. Mining at Keld Heads can be traced back to the 12th century, making it the oldest mine in the dale, and one of the oldest in the orefield. Its hey-day was, however, during

the 19th century when it employed 250 men and boys. It closed in 1888. Green crystals of smithsonite from Chaytor Rake are in the Russell collection.

Wharfedale, Grassington Moor and Greenhow

Although lead mining has made its mark on the hills that flank Upper Wharfedale, veins are generally much less common here than in Swaledale to the north, or Grassington and Greenhow to the south, and the associated smelt mills were generally small. Mines were worked in the Starbotton, Kettlewell and Conistone areas, where remains of the old workings may still be seen.

High on the western side of Buckden Pike, east of Buckden village, lie the workings of Buckden Gavel mine, with further north, on the same vein system, the Bishopdale Gavel mine. Old bell pits in the area suggest very old working, but documentary evidence of mining here seems to date back to the early years of the 18th century when lead ore is known to have been smelted

GRASSINGTON HIGH MOOR LEAD VEIN BLOCK, NORTH YORKSHIRE

The authors (Bob Symes on the left, Brian Young on the right) in the Geological Museum with a large section of a lead vein from Grassington High Moor. The specimen consists of repeated bands of galena and barite. It was presented by the Duke of Devonshire to the Great Exhibition of 1851 and was later displayed in the Geological Museum for many years. The vein section is up to 35 cm across. (©NHMPL)

GRASSINGTON MOOR FLUE, GRASSINGTON MOOR, NORTH YORKSHIRE

Old smelt mill flue with the chimney in the distance. Long horizontal flue leading to chimneys on hilltops were very characteristic of northern Pennines lead smelt mills. The long flues were built to condense lead from the fumes: lead deposited on the walls of the flue was scraped off, mainly by small boys employed in the mills. (©NHMPL)

locally. Mining at Buckden Gavel ended in 1877. Veins and associated flats were worked here. Fine specimens of crystallised hemimorphite are present at Buckden Gavel, and in recent years some examples of baryto-calcite have been found at Bishopdale Gavel.

West of Conistone, on the fellside of High Mark, galena has been worked from a remarkable area of very closely spaced small veins that crop out in the limestone pavements and scars. Associated minerals include sphalerite, barite and fluorite, with rare traces of cinnabar. Although the presence of prehnite has been suggested here, no specimens of the mineral are known and its occurrence here seems very unlikely. It is conceivable that the mineral in question may have been smithsonite or hemimorphite, or even perhaps fraipontite; some of the fraipontite from Virgin Moss mine bears a slight resemblance to prehnite.

Specimens of gold, said to have been panned from Kettlewell Beck, are preserved in the Upper Wharfe-dale Museum at Grassington.

The wide and bleak expanse of Grassington Moor, which lies north-east of Grassington village, contains a great number of very closely spaced veins with associated mine-workings. The earliest definite records of mining here date from 1603 with the discovery of the deposits at Yarnbury. At this time miners were brought from Derbyshire by the Earl of Burlington to develop the mines. His successors, the Cavendish family, Dukes of Devonshire, still hold these and many of the Derbyshire mines. Such was the influence of these miners that by 1640 a series of mining laws and customs, similar to those then in force in Derbyshire, were being applied on Grassington Moor. Mining by small partnerships continued through the 17th century, but in the 18th century an influx of capital from wealthy entrepreneurs encouraged the greater development of the deposits, with the sinking of deeper shafts and the installation of up-to-date pumping and winding equipment. A new smelt mill was built in 1750. Seriously declining outputs from the mines towards the end of the century prompted the Duke of Devonshire to invest directly in a major development of the field. This took the form of the driving of a deep adit system, known as the Dukes Level, under the direction of two successive chief agents: Cornelius Flint from 1779 and John Taylor from 1818. The scheme proved successful and was to sustain the mines until 1876. During the 20th century some short-lived attempts were made to re-work parts of the spoil heaps for fluorspar and barite, the last as recently as the 1960s.

In the 1950s the Grassington Moor and Greenhow area mines attracted the attention of A. W. G. Kingsbury, whose collection, now in the Natural History Museum, London, includes examples of such rare species as minium, plumbojarosite, argentojarosite, parahopeite, spencerite, cinnabar, metacinnabarite,

DRESSING PLANT AT HARRIS SHAFT, NORTH RAKE MINE, GREENHOW HILL, NORTH YORKSHIRE, EARLY 20TH CENTURY

On the left is a set of hotching tubs, and on the right hand crushing is in progress. This picture, thought to have been taken in the early 1900s, has been carefully posed for the photographer. (Beamish)

calomel and native mercury. With the exception of minium and cinnabar, none of these minerals have been found by more recent collectors.

The tiny village of Greenhow, which lies between Grassington and Pateley Bridge, stands at the centre of a small but dense concentration of veins. The area has a very long history of lead mining, with claims that Iron Age, or even Bronze Age, mining may have taken place here. Pigs of lead with Roman inscriptions, found near Greenhow, strongly suggest Roman mining and smelting, but the earliest documentary proof of mineral-working dates from the 12th century. Incomplete records of mining in the area during the 16th and 17th centuries are known, but, as in most parts of the northern Pennines, it was the 18th and 19th centuries that were to see the peak years of lead production. The area contains several sites of early industrial archaeological significance. Amongst these are the Jackass Level, a narrow hand-cut tunnel in hard gritstone, possibly dating from the 17th century. Nearby, the Sam Oon Level is a small passage, perhaps driven as a very early drainage level, although its date and purpose are still matters of dispute. The curious gritstone boulder, known as Sam's Party Oon, may have been an early lead ore grinding mill. During the 20th century small amounts of fluorspar and barite have been produced.

Today at Greenhow the remains of several old dumps and levels may still be seen, together with the fragmentary remains of the Cockhill Smelt Mill to the north of the village. Three separate branches of Greenhow Rake, one of the district's most prominent veins, remain exposed in the long-abandoned Greenhow, or Duck Street, Quarry, south of the village. Two assays of galena from the Jamie vein at Greenhow were said by the former Greenhaugh Mining Company to have revealed 3 and 7.42 dwt/ton (4.2 and 10.3 ppm) gold, although modern analyses have failed to support these claims.

Mineral extraction at Greenhow today takes the form of large-scale limestone quarrying. Coldstones Quarry, east of the village, has cut a number of veins formerly worked by the lead miners. Over the past few years the quarry has become celebrated for good specimens of a number of minerals. These include fluorite, commonly as colourless cubes with very distinctive deep purple outer edges, 'cock's-comb' barite, well-

GREENHOW RAKE, GREENHOW (DUCK STREET) QUARRY

Greenhow Rake, one of the major veins of the Greenhow area, splits into three branches here in Greenhow, or Duck Street, Quarry. The widest of these, the middle branch, is seen here forming a prominent upstanding rib. Old workings for lead and fluorspar can be seen within the vein. (©NHMPL)

formed crystals of hemimorphite, cerussite, smithsonite and some striking examples of vivid red earthy cinnabar. The first British occurrences of the rare minerals prosopite, doyleite and otavite have been reported from here.

A mile or so north of Greenhow, in Ashfold Side Beck, are the dumps of the Prosperous-Providence mine, a relatively late 19th-century lead mine from which a number of specimens of well-crystallised white strontianite have been obtained. Good green pyromorphite has been found in the Merryfield vein. The most recent working for vein minerals in this area, in this case for fluorspar, ended in 1967 at Burhill mines near Skyreholme. Some of the fluorite from here was of a very deep purple tint, somewhat reminiscent of Derbyshire 'Blue John'.

The West Cumbrian Iron Orefield

The hematite deposits of the west Cumbrian iron ore-field occur mainly in Lower Carboniferous limestones within a comparatively narrow belt of country between the villages of Lamplugh and Calder Bridge. Most of the mines are situated around Frizington, Cleator Moor, Cleator, Bigrigg and Egremont. The orefield may be considered in two parts. Between Lamplugh and Egremont the Carboniferous limestones, and their contained hematite deposits, are exposed at the surface, except where concealed beneath glacial and later deposits. From Egremont to Calder Bridge the lime-stones are concealed beneath an unconformable cover of Permo-Triassic rocks. In addition to these deposits, a number of scattered hematite veins within the Lower Palaeozoic rocks of the adjoining Lake District comprise an outlying portion of the orefield.

Small heaps of bloomery slag, found near known hematite deposits, indicate iron working, possibly in Roman times or even earlier. The earliest documentary record of iron mining in Cumbria relates to the grant-ing of a mine in the Parish of Egremont, probably at Bigrigg, to the Abbey of Holme Cultram in 1179. Iron ore is known to have been worked at Bigrigg, more or less continuously from 1635 to 1701. In 1694 ore was being sent by boat from Whitehaven to the Forest of Dean for smelting. Around 1700, iron ore, mined in the Langdale and Coniston fells, was being smelted at a furnace in Langdale. By 1747 hematite, mined at Yeathouse near Frizington, was being shipped from Parton near Whitehaven. New furnaces at Maryport and Seaton were smelting locally-mined kidney ore between 1750 and 1752; and between 1753 and 1755 hematite was being mined at Egremont and Bigrigg. Iron mining is also known at Cleator Moor in 1782, and it is recorded that over 20,000 tonnes of hematite per year were being sent from here to the Corran foundry in western Scotland in 1790-91.

The middle years of the 19th century were to see the beginning of large-scale iron ore mining and smelt-ing in west Cumbria. The first hematite was raised from Woodend and Cleator Moor mines in 1846, with mining starting at Crossfield in about 1860, soon followed by Montreal mine in about 1862. In the adjoining Lake District, hematite mining began at Knockmurton in 1853 with the small cluster of mines around Boot in Eskdale starting in earnest in the 1870s. Many of these mines had closed, or their output had fallen drastically, by the beginning of the 20th century. By the 1920s only a handful of these older mines remained active, although famous mines such as Beckermet and Florence were still comparatively new with their peak years still ahead. It was not until 1939 that one of the orefield's last surviving mines, at Haile Moor, opened.

As the most accessible ore in the northern, exposed, portion of the orefield became exhausted, it fell to the newer mines in the concealed orefield south of Egre-mont to maintain the area's output for much of the 20th century. Large-scale mining and associated exploration for new reserves continued into the 1970s, although by now the industry was in terminal decline. With the closure of the combined Beckermet-Florence-Ullcoats mine in 1981, a major industry finally came to an end. A small company, formed to extract ore from a shallow portion of the orebody at Florence mine, continued to employ a few miners until 2007, but the output of ore was meagre when compared with the peak years of the industry.

Exploration for, and exploitation of, the west Cumbrian orebodies has resulted in the sinking of hundreds of boreholes and shafts. Some confusion can arise over the naming of many of these, with the same shaft commonly acquiring two or more names as mines amalgamated with their neighbours. The positions of a selection of shafts and mine sites is shown on the six-inch scale County Series geological maps of the British Geological Survey which were surveyed in the late 1920s. Recent years have seen the obliteration of many of the surviving features of iron ore mining as build-ings have been demolished and spoil heaps removed: little remains to mark the site of many once famous mines.

The Knockmurton mines at the north eastern extremity of the orefield worked a series of fissure veins of hematite up to 7 m wide within the Skiddaw Group mudstones. In places the ore was followed to a depth of at least 180 m, below which the veins generally became too narrow or poor to be worked. Much of the ore consisted of kidney ore, commonly composed

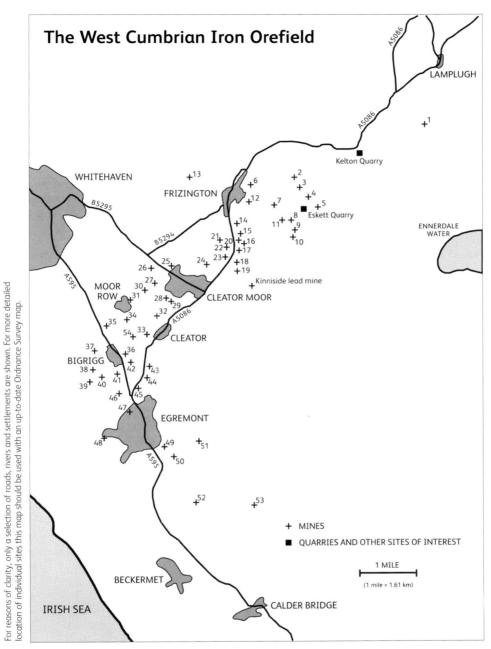

The West Cumbrian Iron Orefield

LAMPLUGH

Kelton Quarry

WHITEHAVEN

FRIZINGTON

Eskett Quarry

ENNERDALE WATER

Kinniside lead mine

MOOR ROW

CLEATOR MOOR

CLEATOR

BIGRIGG

EGREMONT

+ MINES

■ QUARRIES AND OTHER SITES OF INTEREST

1 MILE

(1 mile = 1.61 km)

BECKERMET

CALDER BRIDGE

IRISH SEA

For reasons of clarity, only a selection of roads, rivers and settlements are shown. For more detailed location of individual sites this map should be used with an up-to-date Ordnance Survey map.

1	Knockmurton & Kelton mines	15	Frizington Park No. 14 mine
2	Parkside Winder mine	16	Frizington Park No. 1 mine
3	Windergill mine	17	Goose Green mine
4	Postlethwaite's Eskett mine	18	High House mine
5	Salterhall mine	19	Holebeck mine
6	Lonsdale mine	20	New Parkside mine
7	Margaret mine	21	High Crossgill mine
8	Eskett No. 9 mine	22	Crossgill mine
9	Eskett Park mine	23	Parkside mine
10	Eskett No. 5 mine	24	Birks mine
11	Agnes mine	25	Berrier mine
12	Yeathouse mine	26	Crowgarth mine
13	Millyeat mine	27	Crossfield mine
14	Mowbray mine	28	Jacktrees mine

29	Todholes mine		Wyndham Pit
30	Montreal mine	42	Woodend mine
31	Dalzell's Moor Row mine	43	Longlands mine
32	Cleator mine	44	Rowfoot mine
33	Cleator Glebe mine	45	Townhead mine
34	Postlethwaite's Moor Row mine	46	New Clints mine
35	Sir John Walsh mine	47	Gillfoot Park mine
36	Moss Bay mine	48	Helder mine
37	Parkhouse mine	49	Ullbank mine
38	Pallaflat mine	50	Florence mine
39	Southam mine	51	Ullcoats mine
40	Sike House mine	52	Beckermet mine
41	Bigrigg mine, including	53	Haile Moor mine
		54	Gutterby mine

of small mammillated surfaces. The mines were served by a standard gauge railway and worked from about 1853 to 1914. Large dumps of hematite-rich spoil still mark the site of the mines.

Several well-known mines were worked to the east of the village of Frizington. At Lonsdale mine, which first raised ore in 1879, vugs, or 'loughs' as they are usually known in west Cumbria, up to 9 m across, were common in the orebody, although no details of their content were given. Small dumps remain at the Windergill and Eskett mines and have yielded good examples of kidney ore, quartz, including much yellow jasper, and a little barite. The Margaret mine, which worked from about 1854 to 1923, produced a good deal of kidney ore.

The nearby limestone quarries at Kelton and Eskett have long been known as sources of beautiful aragonite crystals, found in cavities in dolomitised limestone adjacent to major joints and faults, and almost certainly genetically related to the hematite mineralisation.

South east of Frizington, Mowbray mine yielded excellent kidney ore and some beautiful specimens of barite crystals. Superb kidney ore was also a feature of the spoil heaps of the nearby Goose Green mine, which also provided fine specimens of barite and dolomite.

The spoil heaps of both of these mines have now been removed and little remains to mark the site of these once famous localities.

Immediately west of Goose Green, a substantial subsidence pond, west of the A5086 Frizington to Cleator Moor road, marks the position of some of the workings of the Parkside mines. During the working life of these mines, some exceptionally fine crystals of yellow, pale blue and pale green barite were recovered, notably from the pit known as Dalmellington mine.

South of Frizington the small town of Cleator Moor developed rapidly as a mining settlement during the middle years of the 19th century. In the early 1840s Cleator parish had a population of only 750. By 1880 this had swollen to over 10,000, largely due to the arrival of immigrant workers, mainly from Ireland, to work in the expanding coal and iron ore mines. Cleator Moor stands on the western edge of the iron orefield where the hematite-bearing limestones dip beneath the Coal Measures. The town was thus ideally placed to exploit iron ore, limestone and coal to feed the blast furnaces which were built here in 1842. Crowgarth mine, near the western end of the town, probably started in the 1780s, but by the early years of the 19th century it was believed to be exhausted and was closed. It was reopened in 1840 and worked from several

CUMBRIAN HEMATITE

Typical 'kidney ore' about 14cm across, from Cleator Moor [NHM spec. no. BM 28265].

Group of grey metallic conical masses of 'pencil ore' 9cm across, from Cleator Moor [NHM spec. no. BM 43555], purchased from Mr Wright, June 1870.

Two slender tapering rods of fine 'pencil ore' from Furness, Cumbria [NHM spec. no. BM 1960, 447], presented by A. W. G. Kingsbury in October 1960. (©NHMPL)

shafts, including the St John's and York pits. During this time it raised excellent quality hematite, including a considerable amount of kidney ore and 'pencil ore'. It is recorded that many tonnes of the best quality pencil ore from York Pit were sold to the jewellery trade. Fine specimens of specular hematite, quartz, calcite and barite were also a feature of this mine, and lenses of chalcopyrite, largely altered to malachite, were also found in parts of the orebody. Ironically Crowgarth mine, which contributed so much to the prosperity of the town, was also a potent agent in its destruction. By the beginning of the 20th century it was apparent that most of the mine's ore reserves lay at comparatively shallow depth beneath parts of the town. In order to allow extraction, several buildings were demolished and roads diverted. Although some parts of the town were initially spared from undermining, with the severe unemployment problems of the 1920s the local council eventually allowed extraction in order to keep the mine in production.

Montreal mines, which lay south-west of Cleator Moor, were first worked in 1861 and closed down in 1925. Some ore was originally worked opencast. Kidney ore was abundant in parts of the orebody, and in places loughs up to almost one metre across occurred, lined with 'spar' – presumably quartz and calcite. There were several shafts, from one of which, the No. 4 Shaft, coal as well as hematite was raised.

Moor Row is a good example of a west Cumbrian mining village with terraces of small cottages. During the 1980s Gutterby mine, one of many mines between here and Bigrigg, was accessed by a group of local mine explorers who recovered numerous fine examples of bright sky-blue fluorite from loughs within the orebody.

Several mines, famous for the specimens recovered from them, surround the village of Bigrigg. As has already been noted, Bigrigg may be the earliest known centre for hematite mining in west Cumbria, although the site of the early workings cannot now be identified. Most of the famous mines here came into production during the 19th century and, with the exception of the Wyndham or Langhorn Pit, which continued to work until 1938, all were abandoned in the first decade of the 20th century. At Wyndham Pit manganese ore was found as pockets up to a few metres in diameter in the limestone on the margins of the orebody. In places the

proportion of manganese ore to hematite was said to be about 40 or 50 tonnes of manganese ore to 1000 tonnes of hematite. The manganese ore was recovered and sold both to the Workington steelworks and to potteries in Staffordshire. Manganese minerals here included pyrolusite, braunite, manganite and perhaps traces of ramsdellite, although the site is perhaps best known to collectors for specimens of well-crystallised hausmannite. Sir Arthur Russell collected a few striking examples of rose pink rhodochrosite here in the 1930s. Working of the comparatively shallow orebodies at Bigrigg produced a considerable surface collapse, around which the A595 Bigrigg to Egremont road is forced to make a sharp detour.

Most spectacular of the many shallow subsidence hollows to the south east of Bigrigg is the large pond today known as Longlands Lake alongside the River Ehen above the workings of Longlands mine. Ore was first raised here in 1879 from a very shallow orebody which cropped out beneath the alluvial gravel. Elaborate precautions, including restriction of working to one shift per day, and the diversion of the river, had to be taken to allow mining. The cost of timber supports was extremely high and in places alluvial gravel was allowed to run into the workings to aid support. Despite these measures, flooding was a persistent problem and the mine was abandoned in 1924. The same problems resulted in a serious accident at the nearby Townhead mine in 1919. Three miners were trapped underground by a sudden inrush of water. One man died, but his colleagues survived by reaching the foot of a borehole from the surface where they were supplied with air, food and drink for five and a half days until pumping lowered the water level sufficiently for them to be rescued.

South-west of Bigrigg the Pallaflat mines, including Southam and Sike House pits, are famous for magnificent specimens of calcite. A variety of forms were present, but the combinations of scalenohedron and rhombohedra, the so-called 'butterfly twins', are perhaps the best known. Many were transparent, but inclusions of black manganese oxides were common. Mining at Pallaflat ended in 1914 and the few remaining dumps give meagre evidence of this once famous group of mines. In the 1980s excavation of a small dump from the Southam workings, now completely

WYDHAM PIT, BIGRIGG, CUMBRIA

The ruins of the engine house at Wydham Pit, now demolished. This mine, started in 1867 by Lord Leconfield, was a noted source of fine crystals of calcite as well as well-crystallised hausmannite and rhodochrosite. (©NHMPL)

ADVERTISEMENT FOR FLORENCE MINE FROM THE 1980s (©NHMPL)

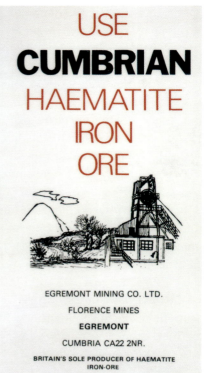

USE
CUMBRIAN
HAEMATITE
IRON
ORE

EGREMONT MINING CO. LTD.

FLORENCE MINES

EGREMONT

CUMBRIA CA22 2NR.

BRITAIN'S SOLE PRODUCER OF HAEMATITE
IRON-ORE

removed, yielded a number of very fine examples of typical Pallaflat calcite. Very similar calcite specimens are well known from the nearby Gillfoot Park and Helder mines which closed in 1924, the sites of which are now built over.

The handful of mines south of Egremont, which were the last to be worked in the orefield, exploited very large deposits of hematite within the Carboniferous limestones concealed beneath a cover of Permo-Triassic rocks.

Ullcoats mine began production in 1900 and continued working until 1968 when Florence mine, with which it was then connected, also closed. Ullcoats produced good specimens of all hematite varieties, together with excellent examples of the typical gangue minerals, including fine specimens of pale blue fluorite. Perhaps best known for spectacular fluorite specimens was the adjacent Florence mine. Sinking of this mine began in 1914 with the first ore being reached in 1920. The Florence-Ullcoats orebody proved to be the largest known in Cumbria and, together with extensions southwards, these huge deposits were to sustain the industry throughout much of the 20th century. A feature of the Florence orebody was the abundance of

cavities or loughs lined with well-crystallised calcite, dolomite, quartz, kidney ore and specular hematite. Kidney ore was especially common at Florence and the mine soon earned a reputation for fine specimens. Although most loughs were around 0.3 m across, much larger cavities up to 4 m or more in height were also found. Unfortunately for the mining company, the original Florence Shaft, the No. 1 Shaft, was sunk directly through the orebody. By the 1940s it was necessary, if the mine was to continue in production, to sink a new shaft outside the orebody in order to allow extraction of the large reserves remaining within the No. 1 Shaft pillar. Florence No. 2 Shaft was sunk in 1947 and served as the main access to the mine until the 1980s.

To the south of Florence, the No. 1 Shaft of Beckermet mine was sunk in 1903 with the first ore raised in 1906. The No. 2 Shaft, also known as Winscales Pit, was sunk between 1916 and 1918. Beckermet mine became a very extensive mine with workings extending nearly 1.5 miles southwards, almost to beneath Calder Bridge. Mineralisation was not continuous, but occurred as a series of orebodies associated with a number of large NW-SE trending faults. In its latter years the

LONELY HEARTS OREBODY, FLORENCE MINE, EGREMONT, CUMBRIA

Workings in the Lonely Hearts Orebody, Florence Mine. (©NHMPL)

FLORENCE MINE, No. 2 SHAFT, EGREMONT, CUMBRIA

Headframe and surface buildings at Florence Mine, No 2 Shaft. (R. F. Symes)

mine was connected with the Florence-Ullcoats workings to the north and became the last surviving major hematite mine in west Cumbria.

The last hematite mine to be sunk in west Cumbria lay about one mile east of Beckermet at Haile Moor. A single shaft was sunk here in 1939 to work a group of large orebodies associated with a complex series of faults. Although an underground connection with Beckermet was contemplated, it was never driven. Ore from Haile Moor was conveyed by aerial ropeway to a loading bay on the railhead at Beckermet. The mine closed in 1973.

Both Haile Moor and Beckermet mines were well-known sources of specimens of most of the main minerals of the orefield. In the latter years of mining, all specimens were rigorously regarded as property of the company and were marketed directly to dealers from the company's offices at Workington.

Shortly after the closure of the combined Beckermet-Florence-Ullcoats workings in 1980, a small company was formed by a handful of the original workforce to resume small-scale mining. Their target was a shallow portion of the orebody, known as the Lonely Hearts orebody, which had been partially worked but never completely extracted. Mining continued here until 2007, but compared to former years

the output was tiny. The Florence No. 2 Shaft was abandoned as the main access, and replaced by a newly driven inclined adit from the surface. Ore has been supplied to specialised steel makers in Europe and to the makers of red pigment. Fine specimens of hematite, together with some blue fluorite, were recovered from time to time.

The South Cumbrian Iron Orefield

Lying to the south of the Lake District, the south Cumbrian iron orefield is separated by the estuary of the River Duddon into two unequal parts, although it has been suggested that as yet unknown orebodies may lie beneath the estuary. West of the Duddon estuary, the orefield is centred around the small town of Millom, site of the famous Hodbarrow mine and the former Millom steelworks. East of the Duddon estuary, in the area known as Furness, the most extensive portion of the orefield is centred around the market town of Dalton-in-Furness.

As in west Cumbria the hematite deposits of south Cumbria are mainly hosted by Lower Carboniferous limestones. The only deposit known to have been worked in host rocks of a different age is that of the Waterblean, or Waterbleau, mine near Millom, where hematite occurred in limestones belonging to the Ordovician Dent Group. In common with those of west Cumbria, the deposits of south Cumbria include vein-like bodies developed along faults and large replacements of limestone. However, unique to the Furness area is another group of deposits known as 'sops'.

South Cumbria has enjoyed a long history of hematite mining. Stone hand-axes of presumed Neolithic age, found beside a face of hematite in an old working near Stainton, may give evidence of pre-Roman working. An account in the records of Furness Abbey of a dispute between the Abbot of Furness Abbey and Henry of Orgrave in 1235, about iron ore from the Dalton area, is the first documentary proof of hematite working. Iron ore mining is also recorded from the Mouzell area, near Dalton, as early as 1271. A grant to work iron ore at Plumpton was made to Conishead Priory in the reign of Edward II (1307-27) and references are known to iron ore mines at Dalton, Orgrave and Marton in 1400. By the dissolution of the monasteries in 1537, smelting of local iron ore with charcoal had led to a shortage of timber. Smelting with charcoal is indicated by a furnace on the River Duddon on John Speed's map of 1745, and four furnaces are known to have been in operation in Furness in 1750. In the Millom area iron ore mining was already in progress in 1777.

However, as in west Cumbria, it was the 19th century that was to witness the development of iron ore mining and smelting that was to make such an indelible mark on the landscape and economic history of south Cumbria.

Trial workings near Millom in the 1850s were to blossom into the orefield's largest, most productive and latest surviving mine at Hodbarrow. Initially an extensive flat-lying body of ore was discovered and worked north-westwards towards Millom from the Old mine. Further exploration revealed southward extensions of the ore, together with two further separate orebodies at shallow depth beneath the coast. In order to work these safely, a novel mining method was devised. Known as 'sand filling', this involved filling worked-out spaces with sand, pumped into the workings as a slurry. At the surface a long curved sea wall, known as the Inner Barrier, was constructed in 1905 to protect the workings from flooding. However, as the workings followed the ore further offshore, this collapsed and a new barrier, the Outer Barrier, had to be built. Since the mine closed in 1968 all surface buildings have been demolished and the numerous shafts sealed. The Outer Barrier and the flooded ground on its landward side, now developed as a yachting marina, comprise one of the most spectacular legacies of the mining industry anywhere in northern England.

Much of the Hodbarrow ore consisted of 'hard blue' massive hematite, although some fine examples of large kidney ore were found, particularly in the Old mine. A curious feature seen in several old specimens from here is the presence in the mammillated surfaces of pseudomorphs after a cubic mineral, probably pyrite or fluorite. Calcite crystals were commonly found encrusting the kidney ore, and in places on the margins of the orebody a partially hematised siliceous limestone, known by miners as 'red rat', was found.

On the east side of the Duddon estuary, large dumps at Askham mark the sites of Sandy's and Woodhead mines. The Sandy's orebody was a flat-lying replacement deposit very similar to that at Hodbarrow. The sop worked at Woodhead is said to have contained a little manganese ore near its base.

South of Roanhead were Cumbria's largest sops, formerly worked at the Nigel, Rita, Sandscale, Park and California mines. The largest of these, the Park sop, was discovered in 1849 and worked until 1921. The abundance of ore from these newly-discovered deposits, together with the ready availability of Durham coke which was ideally suited to iron smelting, via the recently opened Furness railway, prompted the founding of the steelworks at Barrow. The Rita sop was found in 1875 and was worked out by 1925. The westernmost of these deposits, the Nigel sop, was discovered in 1902 and remained in production until 1950. Fine examples of kidney ore were recovered from these mines, especially from Nigel Pit where, in addition, some smoky quartz was also found.

To win ore from the sops a system of mining, known as 'top-slicing', was employed. In this method the roof was allowed to collapse as progressively lower slices of ore were removed. This resulted in the development of huge conical depressions at the surface which flooded

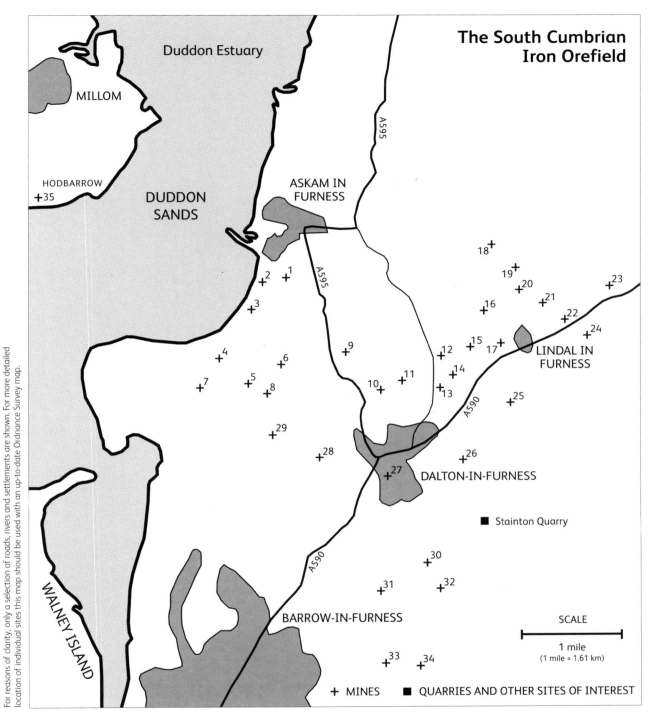

The South Cumbrian Iron Orefield

Duddon Estuary

MILLOM

HODBARROW
+35

DUDDON SANDS

ASKAM IN FURNESS

A595

A595

18+

19+
20+

23+

16+

21+

22+

24+

2+ 1+

3+

9+

15+ 17+

LINDAL IN FURNESS

4+

6+

12+

7+

5+
8+

10+ 11+

14+

13+

25+

29+

28+

27+

26+

DALTON-IN-FURNESS

A590

■ Stainton Quarry

30+

A590

31+ 32+

BARROW-IN-FURNESS

WALNEY ISLAND

SCALE

1 mile
(1 mile = 1.61 km)

33+ 34+

+ MINES ■ QUARRIES AND OTHER SITES OF INTEREST

For reasons of clarity, only a selection of roads, rivers and settlements are shown. For more detailed location of individual sites this map should be used with an up-to-date Ordnance Survey map.

1 Greenscoe mine	10 Elliscales mine	19 Carkettle mine	28 Goldmire mine
2 Sandys mine	11 Mouzell mine	20 Lindal Moor mines	29 Thwaite Flat mine
3 Woodhead mine	12 Tytup mine	21 Whinfield mine	30 Newton mine
4 Nigel mine	13 Crossgates mine	22 Diamond Pit	31 West Newton mine
5 Roanhead mine	14 Eure Pits mines	23 Pennington mine	32 North Stank mine
6 Park mines	15 High Crossgates mine	24 Lowfield mine	33 Yarlside mine
7 Sandscale mine	16 Whitriggs mine	25 Lindal Cote mine	34 Stank mines
8 California mine	17 Black Guards mine	26 Dalton mine	35 Hodbarrow mine
9 Green Haume mine	18 Poaka mine	27 Anty Cross mine	

rapidly when pumping of the workings stopped. The subsidence ponds which are conspicuous landscape features, together with a few overgrown spoil heaps, are all that remain today to mark the site of these once large and productive mines.

Sops were also numerous in the area immediately north and north-east of Dalton-in-Furness and, although none compared in size to those of the Park mines, they proved very productive at the Elliscales, Mouzell, Crossgates and Eure Pits mines. Several large subsidence craters today mark the sites of some of these workings.

Immediately to the north-east lay the major group of mines which worked an extensive series of replacement and vein-like orebodies associated with the NW-SE trending Lindal Moor fault system. The most prominent of these ore-bearing structures was known as the Lindal Moor vein, although in detail this was a complex structure composed of a plexus of ore-bearing faults. The heyday of mining on Lindal Moor was during the 19th century, although several small-scale operations continued until the early 1930s. A vain attempt at reviving mining on Lindal Moor was begun towards the end of the Second World War with the development of the small Margaret mine. However, this venture was not successful and, after a change in ownership in 1959, when an abortive attempt was made to win ore by opencast working, the mine, the last to work in Furness, finally closed. The area is today marked by extensive areas of overgrown and collapsed ground, known locally as 'broken ground'. Kidney ore was present locally in the Lindal Moor deposits; good examples of this, suitable for use in jewellery making, were being produced from the Whinfield pits in the early 1920s. One of the Lindal mines, the Diamond pit, possibly owes its name to the local abundance of quartz crystals within loughs in the ore. Well-crystallised dolomite is locally present on the spoil heaps at Lindal Moor, although few specimens have been found to rival those of west Cumbria. Calcite in a variety of forms is also common at Lindal Moor, but pride of place for specimens of this mineral belongs to the Stank group of mines, south of Dalton-in-Furness.

The mines of Newton, West Newton, North Stank, Stank and Yarlside, which worked vein-like orebodies in faults in the Lower Carboniferous limestones, were all developed during the second half of the 19th century. Most had closed before the Second World War,

KATHLEEN PIT, ROANHEAD MINES, ASKAM IN FURNESS, CUMBRIA

The Cornish Pumping Engine House, built by Hodge of St Austell, Cornwall, in 1851. (Dalton Castle Archive)

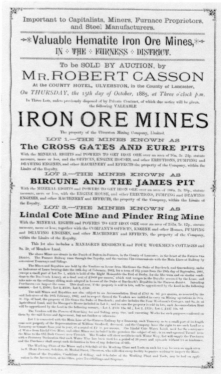

ADVERTISEMENT FOR SALE OF FURNESS IRON ORE MINES, FROM 1895 (©NHMPL)

although all mining finally ended here shortly after the end of the war. Today the grassed-over courses of the railways that served the mines, a number of relatively small spoil heaps, and the ruined walls of the Cornish engine house at the Stank No. 1 pit, are all that remain of these mines. To mineralogists the Stank mines were famous for the many specimens of beautiful calcite crystals recovered from them. Indeed, specimens from here are amongst the finest from any Cumbrian hematite deposit.

Trial-drilling during the First World War revealed a vein-like deposit of hematite on the outskirts of Dalton-in-Furness which was developed at Anty Cross mine. The deposit was remarkable for the occurrence within the hematite of pockets of massive chalcocite and chalcopyrite, with small amounts of pyrite, locally much altered to malachite. A few tonnes of copper ore were worked here alongside the hematite. The copper minerals may have been present prior to the hematite mineralisation. The only other occurrence of significant copper mineralisation in Furness is at Sea Wood, near Bardsea, where a small vein in Lower Carboniferous limestones carries chalcocite, malachite and azurite. This vein is apparently unrelated to the hematite mineralisation.

During the early 1990s the working faces of Stainton Quarry exposed a small hematite vein from which a number of fine specimens of kidney ore were collected. In many of these the characteristic coarsely mammillated hematite encrusts an earlier generation of massive to finely mammillated hematite which clearly pseudomorphs large calcite scalenohedra.

No. 2 PIT, STANK MINE, BARROW-IN-FURNESS, CUMBRIA, *circa* 1870. (Dalton Castle Archive)

24 Dark violet cubic Fluor, sprinkled with white
 nummular 12-sided Spar, brown Rhombic Barytes,
 and small grains of Blende. From Alston Moor,
 Cumberland. —————————————— 0,10,6

25 Aqua-marine and white cubic Fluor, with a beautiful
 opaline surface. From Saxony. ————————— 1,1,0

26 Large cubes of Amethystine Fluor, with brilliant
 double-pointed Quartz Crystals. From the Bishopric of
 Durham. ————————————————— 2,12,6

27 Large Group of Violet cubic Fluor, sprinkled with
 minute Crystals of Mun___ a Bed of white Cauk.
 From Derbyshire. ————————————— 0,10,6

28 White stellate___ ____ ____ ____ the cavity of
 a black rock st___ ————————————— 4,4,0

29 Violet cubic F___ ____ ____ ____ a Cube,
 each of who___ ____ ____ ____ quadrangular
 pyram___ ____ ____ ____ ____tallized
 green___ ____ ____ ____ ____ Marcasites,
 &c. ——————————————————— 3,3,0

30 Striated ____ ____ —————————————— 0,5,0

31 Capilla___ ————————————————— 1,1,0

32 Zeolite___ ____ ____ ____ ____ the
 Hartz.

33 White Z___ ————————————————— 0,15,0

34 A very c___ ————————————————— 1,1,0
 Crystals, on a ___

35 Saphirine F___ ____ ____ ____ ____ Ore, 3,3,0
 Probably from Hun___ ————————————— 1,1,0

36 Red Zeolite, form___ ____ ____ ____ ____
 favouring most the tab___ ____ ____ ____ ___rk cellular
 Basaltic Stone. From ___ ____ ____ ___nia. R. R. — 5,5,0

Collectors
and Collections

THE north of England has a strong and long tradition of mineral collecting with records dating back to the 17th and 18th centuries. Specimens of northern England minerals are represented in the Woodward collection (1665-1728), and in the Cracherode collection (1730-99). The prominent mineral dealer Jacob Forster (1739-1806) was selling north of England fluorite, sphalerite and barite in his Paris shop in the 1780s. Daniel Boulter of 19 Market Place, Yarmouth was dealing in barite from Dufton about 1793. The catalogue of Boulter's collection lists some 15 pages of minerals, rocks and fossils with some locality detail. No. 74, for example, is a fine specimen of amber-coloured tabular barite from 'Dofton mine, Westmoreland, very scarce'. Specimens of calcite (calcareous spar) from the north of England were figured by Rashleigh (1797; 2nd vol. 1802) and other specimens figured by Sowerby (between 1804-17). Specimens from the Victorian era of mining are common in major museum and private collections.

Judging by the number of north of England specimens represented in old collections, it is clear that specimen-collecting was commonly practised by the miners, who no doubt regarded it as a perk of the job. Certainly both miners and management appreciated the beauty of the 'spar' specimens, though sales to dealers would have needed approval. During the 20th century the management at the Florence-Beckermet mines operated 'an incentive scheme', begun in 1907, to encourage miners to recover and handle with care any crystal finds of obvious quality. However, it is difficult to equate this with the known policy of the Beckermet management in its final years, where miners were compelled to hand over specimens. Failure to do so was a sacking offence.

As early as 1802, writing in the *Beauties of England and Wales*, Britton and Brayley stated:

Beautiful specimens of Spar of various colors [sic], amorphous, and crystallized in different forms, are found in lead mines of Aldston-Moor; and since the study of mineralogy has become fashionable, have been sold for considerable sums.

OPPOSITE

Page of Revd Clayton Mourdant Cracherode's collection catalogue, together with a fine group of purple fluorite cubes (No. 26) from Co. Durham, listed and described on this page. (©NHMPL)

The Lead mines are chiefly in Aldston Moor … in working some of these mines, the miners frequently meet with large breaks in the rock, like grottos [sic], wholly encrusted with the most beautiful spar, which, on entering, has the richest appearance imaginable. The whole cavern, by the light of a candle, reflected from a thousand points, appears as if bespangled with gold, silver and diamonds. These internal openings are generally closed up as soon as found, the spar they contain being a great temptation to the workmen to neglect the service of their employers, as they could obtain more by gathering and selling spar, than by their own business.

Peter Gilmore, a mineral dealer from Alston in the late 1800s, complained in a letter in 1897 to the British Museum (Natural History) that:

Green Fluors is bad to get now as the Masters has closed up the old mine we got them from on account of so many people going and knocking the place to pieces and might have damages to pay for some of the men getting killed!

Generally it would seem that collecting was unpopular with mine owners, although attitudes were different in some areas and individual mines. Perhaps the early specimen production of the mines would have been influenced by 'gentlemen' collectors who were natural historians/scientists of some standing. Such collectors would have set the agenda for the systematic study of minerals. Indeed, there already existed a dealer market in fine specimens from the likes of Jacob Forster (1739-1806).

Collecting the 'Bonny Bits'

A tradition of the northern miners, which may have had its early origin in the influence of the German and other European miners and advisors brought into northern mines, is in the use of mineral specimens as decorative or ornamental displays, grottos, or in spar boxes.

At the Great Exhibition of 1851 in Hyde Park,

London, Thomas Sopwith, chief agent for W. B. Lead, together with other local agents, put together a display of minerals from the northern Pennines. This was of course to show economic potential rather than perfection of crystallinity. However, perhaps second only to the annual display of their prized leeks, was the desire of the north Pennine miners, and local collectors, to exhibit their prized mineral crystals and rock specimens – 'the bonny bits'. It is known that at exhibitions of 'curiosities' in the Weardale area in the 1860s, mineral specimens would be shown. This is not surprising, for at this time there were many working mines from which many fine specimens were being obtained. However, this time of prosperity soon disappeared as mines closed and a period of mining strife took over. Nevertheless, there was still some production, and considerable interest in displaying locally derived mineral specimens. So it was to be that the first programmed exhibition of mineral specimens was held at St John's Chapel, Weardale on Saturday 24 December (Christmas Eve) 1887, and ran for the next 13 days. The exhibition was opened by Walter Beaumont Esq., of the Beaumont mining family. However, Walter was the

ST JOHN'S CHAPEL EXHIBITION CATALOGUE

Title page of the catalogue for the Grand Mineral and Geological Exhibition, held in St John's Chapel, that opened on 24th December 1887. (©NHMPL)

'black sheep' of the family and his support for the show was perhaps not so much an endorsement by the Beaumont family, but more of a signal of Walter's personal support for the miners at a time of economic depression. At the opening, the Weardale Brass Band entertained. Virtually all of the 35 classes were mineralogical or geological specimens, but a few classes were for pictures and other natural history objects. For instance, Class 4 was for the best piece of 'Brangle ore', Class 6 was for the best 'Spar Diamond', Class 7 for the best specimen of Spar, and Class 9 was for the best Spar Case. First prize in these classes was 15 shillings, second prize 7 shillings and 6 pence, third prize 5 shillings, and fourth 2 shillings and 6 pence. Class 7, the best Spar, had 17 entries – it was won by L. Maddison of St John's Chapel, Weardale. Second was a Mr J. Dawson, also of St John's Chapel. In the same exhibition, Class 9, best Spar Case, was won by Mr Rutherford of Nenthead, with Mr J. Dawson of St John's Chapel again second. A silver cup was presented to Mr J. Thompson from Hill Top for the best collection of minerals.

Interestingly, over 100 years later discussion is still active as to the nature of the actual specimens displayed within some of the more exotic exhibition classes. For instance there were separate classes for 'Brangle', 'Steel' and 'Potter's ore'. The general concensus is that 'Brangle ore' was the name given to a fine-grained mixture of galena, fluorite and country rock, which was almost impossible to separate on the washing rake. It would appear that both 'Steel' and 'Potter's' ore were names given to differing forms of galena. Potter's ore was a pure, softer and silver-poor galena used as a base for pottery glazes. It is believed that a considerable amount of this material was sold especially to Holland, where it was used in the earthenware industry. The term 'Steel ore' was applied to a dull grey, hard and fine-grained sheared galena, commonly with a higher silver content. 'Steel ore' appears to have been associated with post-mineralisation movement along the vein and often shows slicken-siding.

The 1887 Weardale Exhibition was fully and eloquently reported by the *Auckland Chronicle* of Friday, 10 January 1887. This report included the full list of class winners. The report recorded that the scientific gentlemen chosen as adjudicators were thanked for their services '… and it showed that all "partyism" had vanished in the desire to bring to light the wealth which God had placed in the bowels of the earth for the united and general good'. This comment was met with loud cheers.

The programme for a subsequent exhibition, opened by the member of parliament for Barnsley on Saturday 17 August 1901, and which lasted for three weeks, held again in St John's Chapel, lists 48 exhibition classes, most of them related to mineralogical specimens. There was also a greater number of craft classes including collections of china, sewing and paintings. This 1901 show appears to be the last such exhibition in the northern Pennines.

Of course an even greater audience would have studied the north of England specimens on show at 'The Great Exhibition of 1851' in Hyde Park, London. Thomas Sopwith, W. B. Lead's chief agent in the northern Pennines, exhibited not only mineral specimens but geological sections and models.

The tradition of an annual competitive mineral exhibition in the northern Pennines was revived in 1992 at Killhope Lead Mining Museum. Although by then mining was at a very low level, competition amongst collectors proved to be as intense as in the past.

Spar Boxes and Grottos

Wages for miners throughout the north of England were poor, but miners, whether staying in 'mine lodging shops' or at home, tried to earn a few extra shillings by craft work. Amongst these crafts grew the tradition of making 'spar boxes'. This pastime concerned the decoration of variously designed and shaped wooden boxes with specimens procured from the local mines, sometimes obtained by purchase or by exchange with fellow workers. Certainly the miners appreciated the beauty and varied colour of the crystal groups found in their day-to-day work, for their homes throughout the northern moors and dales were decorated with crystal groups on the fireplace or window-sill, and in the garden in rockeries, paths and walls. Some miners would have their collections set in these often 'folk-art' home-made

SPAR TOWERS

Like spar boxes these were assembled from choice examples of local minerals, and were usually displayed under a bell glass. The tower on the left (approximately 70 cm high), in the private collection of David and Elizabeth Hacker, is made almost entirely of minerals from the west Cumbrian iron ore mines. The central example, also in the possession of Mr and Mrs Hacker, is made mostly of northern Pennine minerals (approx. 30 cm high). The tower on the right (40 cm high), now in the collection of Killhope Lead Mining Museum, is built mainly of northern Pennine material, though it also features conspicuous examples of west Cumbrian aragonite crystals. As in many spar boxes, it was common to find evidence of exchange of minerals between the northern Pennines and west Cumbria. (©NHMPL)

cabinets or spar boxes. Crystal groups of various sizes and colour, mostly obtained locally, were used in these boxes. However, West Cumberland ore specimens commonly appeared in some of the Weardale spar boxes, but mostly local material was used. Crystals were secured to the bottom, top and sides of the boxes by plaster of Paris or local varieties of cow gum or putty, and the boxes were protected by a glass front. The shape of the spar boxes varied, the most common being likened to the tops of grandfather clocks. Indeed, in many instances they probably were the tops of grandfather clocks. It is interesting to note that a few mineral dealers were also clockmakers. Other combinations of crystal groups were mounted on a wooden base and covered by a glass dome. Such 'spar towers' or arches appear to have been more common in the west Cumbrian iron orefield. These commonly consisted of

smoky quartz, specular hematite, calcite, aragonite, dolomite and barite. The west Cumbrian spar boxes tended to be simpler, sombre and less ornate than those of Weardale, reflecting the greater use of dark minerals such as specular hematite and smoky quartz.

Today, as in the 19th century, these early spar boxes are considered collectable and fetch appreciable sums at auction. For instance, at the auction of 12 July 1972 at Sotheby's, Belgravia, Lot 212 was 'a large west Cumbrian "spar tower" with crystal columns joined at centre and about 2 feet high on a wooden base …'. The locality given was the Crowgarth mine, Cleator Moor.

As time passed, the construction of these spar boxes became larger, and more and more extravagant in detail and design. Such craft masterpieces would have mirrors fitted into the structure and would be lit by candles or bulbs. Other natural history objects, particularly stuffed birds, were incorporated into their design. Other common designs featured mine layouts with small cast or hand-carved figures of miners, ore tubs, and rails within the crystal layout. The northern Pennine boxes would normally contain attractive 'spar' specimens of galena, sphalerite, fluorite, barite and witherite. These miniature grottos would present the minerals as natural growth clusters, as columns or arches, or in some cases as stalagmitic or stalactitic structures. The exterior of the cabinet or box would often be profusely decorated.

There are reports of rare boxes containing Lake

THE EGGLESTONE SPAR BOX

At 225 cm high this is the largest known spar box made in the northern Pennines. Some doubt remains about the identity of its maker and indeed of the date of its creation. However, it is known to have been doing the rounds at local shows in the early years of the 20th century. The complete box is pictured here in a private house in Weardale prior to its acquisition by, and removal to, Killhope Lead Mining Museum where it can be seen today. On the right is a detailed view of a street scene in the interior of the box. (©NHMPL)

District minerals and rocks, but these are not known to be as spectacular in either quality of specimen or design as elsewhere in the north of England.

There is little doubt that many of the larger more detailed 'spar' boxes were made for show rather than domestic ornament. Certainly the Egglestone Spar box, now in Killhope Lead Mining Museum, is known to have been transported to and exhibited at local agricultural shows and village festivals. It is almost certain that some of this folk art was commissioned by the 'gentlemen' of the day. A national collection of spar boxes has now been assembled at Killhope Lead Mining Museum, Weardale, with an accompanying book published by the Museum.

Unlike Cornwall and Devon and some other British mining areas, there appear to be very few recorded 'grottos' incorporating minerals and rocks from the north of England. In fact, no example of a true grotto of this type in the north of England is known to the authors. However, it seems probable that crystalline groups of purple fluorite used in the original construction of the grotto in the landscaped park at Painshill, Surrey, originated in Weardale. This grotto was created in the 1750s by Charles Hamilton, a younger son of Lord Abercorn, and was a major feature of a fine landscaped garden. In the famous chambered Pope's Grotto, Twickenham, built in 1742 and recently surveyed by one of the authors (R.F.S.), reference is made to the use of purple and yellow spar in the roof believed to be from the north of England. Such specimens were described as from Mr Ord's mines in Yorkshire, and pieces of 'white spar' from the Duchess of Cleveland (Raby Castle Estate) are thought to have been quartz or calcite from the Teesdale area.

Perhaps the most interesting example of decorative use of minerals in a public space is the ornamentation of gate posts, and walls of a house and shop in Keswick in the Lake District.

It is not clear where or when the spar box tradition came into being in northern England. It could have been the influence of miners descriptions of vugs lined with crystals, found in many of the orebodies encountered in the northern England mines, or it could have been a reflection of the early grottos built on fine estates and gardens. However, a major influence may have come from the traditions of European metalliferous miners. Somewhat similar boxes and structures of crystal groups were traditional crafts in the German and central European mining fields from an early date (15th to 16th century), and it may be due to the influence of the German miners brought to work the ores of northern England, especially the Caldbeck Fells in the 16th and later centuries, although there is no evidence of such a tradition among the Caldbeck miners. However, the style of the northern England spar boxes is somewhat different to the more ornate spar ornaments, known as 'handstein', of Germany and central Europe. It could be that they were constructed as miniature versions of the shell and mineral-lined grottos, popular features of continental estates that figured in the popular press of the time.

From the written evidence of the north Pennine spar box makers, it would seem that the key time was the latter part of the 19th century. As mining diminished, the quest for fine mineral specimens became more focused on fewer specimens, but the desire to create mineral folk art remained. This continues today with a modern enthusiasm to build ornamental spar boxes or towers.

GROTTO IN KESWICK, CUMBRIA

A rare example of a grotto adorned with local minerals in Keswick, Cumbria. (©NHMPL)

Collectors and Collections

Some of the important collectors and collections of British Minerals have been detailed in books such as *Minerals of Cornwall and Devon* (Embrey and Symes, 1987), *Minerals of the English Lake District, Caldbeck Fells* (Cooper & Stanley, 1990), *A mineralogy of Wales* (Bevins, 1994) and *Minerals of Scotland* (Livingstone, 2002). Details follow of some important collectors and collections relating to the area described in this volume. The authors are grateful to Michael Cooper of Nottingham Castle Museum and Art Gallery for some of the information on mineral collectors and dealers.

There were probably many early collections of fine northern England specimens, but most have not survived. The collectors and collections discussed here are important in mineralogical and collection history.

John Woodward FRS (1665-1728)

Woodward practised as a physician for more than 30 years. He also made an important contribution to botanical science and archaeology, but it is as a geologist that he is chiefly remembered. He can justly be remembered as the foremost British geologist of the period preceding Hutton, Smith and Lyell, and his name is perpetuated in the Woodwardian Chair of Geology in the University of Cambridge, for the foundation of which he left provision in his will. His interest in geology is said to have been aroused by chance on an invited visit to the home of Sir Ralph Dutton at Sherborne in Gloucestershire. Here he was able to collect objects of natural history and to see exposures of richly fossiliferous Jurassic rocks. In 1692, at only 28 years of age, he was made Professor of Physic at Gresham College, and in 1693 he was elected a Fellow of the Royal Society. In 1695 Woodward wrote his famous book, *An essay towards a natural history of the Earth*. This book contained a theory of the origin of the rocks of the Earth's crust, in the light of contemporary knowledge. In succeeding years Woodward gathered together, at Gresham College, a very large collection of fossils, minerals and other curiosities. He described all his specimens carefully, with information on locality and nature of occurrence. Even by modern standards the collection was extensive and well curated. It included not only specimens of British minerals and fossils, but also examples from Europe, North America and Asia. The collection was probably mostly put together by exchange and donation, but Woodward also collected much of the British material himself. In Woodward's time chemistry could provide little help in identification and systematics, and crystallography almost none for identification, so Woodward's mineralogy bears little resemblance to modern text books. However, Woodward was probably the first British author to publish a work solely devoted to the subject. Woodward's last book, published posthumously in 1729, was a catalogue of his geological and mineralogical collections entitled *An attempt towards a natural history of the fossils of England, etc.* .... This remarkable early collection, which includes some 9000 catalogued entries, is to be found in the Sedgwick Museum of Earth Sciences, Cambridge. It is stored in the original five double cabinets, veneered in burred walnut and in exceptional condition. Historically important for this study are the many catalogue entries of northern England specimens. Listed below is a selection of these catalogue entries. For example:

1. *m16, p.215: Colonel Burley's mine on Richmond Moor, 2m west of Richmond, Yorkshire (believed to be galena).*
2. *m24: Lead ore in perpendicular fissure. Thorngill lead mine about a mile from Austin, Northumberland.*
3. *m66: Entry for Blagill lead mine.*
4. *Mr Bathe's lead mine, on Moulderside Hill, Arkendale, Yorkshire.*
5. *Spars and crystals shot into cubic form Hartley Castle lead mine, Westmoreland.*
6. *Woodward, f.91, p.157: Cubic spar with a cast of purple and some grains of yellow sulphurous marcasite, lead mine at Nenthead in Cumberland.*

The collection also contains a suite of interesting localised fluorites:

- *Fluorite, colourless cubes: Sir Charles Musgrave's lead mines near Hartley Castle, Kirkby Stephen, Westmorland, tome 1, part 1, p. 157, ff. 88-90. The old Hartley mines were situated at Hartley*

Birkett, one mile east of Hartley Castle.
- *Fluorite, a violet cube: Red Groves mine, Middleton-in-Teesdale, Co. Durham, tome 1, part 1, p. 158, f. 92.*
- *Fluorite, violet cubes: Yew Tree Grove, Stanhope, Weardale, Co. Durham, tome 11, p. 78, I.S.*
- *Fluorite, violet cubes: Cross Fell mine, Alston, Cumberland, tome 1, part 1, p. 158, f. 92.*

Philip Rashleigh (1729-1811)

Rashleigh was one of the most important British mineral collectors of the 18th century. Although best known for the superb collection of Cornish minerals, the Rashleigh collection also contains excellent northern England specimens, some of the fluorites being among the finest known. Rashleigh was the eldest son of Jonathan Rashleigh, Member of Parliament for Fowey, Cornwall. Following his father's death in 1764, Philip Rashleigh was elected to his father's parliamentary seat which he held until 1802. It is believed that his interest in minerals began about 1765, and his collections and studies of Cornish minerals enabled him to be elected a Fellow of the Royal Society (FRS). He was also a Fellow of the Society of Antiquaries (FSA). The family home and collection was at Menabilly near St Austell, Cornwall, and it was from here that he received visitors to the collection and corresponded with mineralogists throughout Europe. The greater part of the Rashleigh collection is exhibited or stored at the Royal Cornwall Museum, Truro.

It is known that Rashleigh visited Jacob Forster, the mineral dealer at Covent Garden, London. He was also a customer of Mrs Francis Powell, another mineral dealer in London. It was from here that Rashleigh acquired many of his northern England specimens.

Mrs Francis Powell

Mrs Powell seems to be almost the only female mineral dealer of any consequence who did not take up the trade from her husband following his demise or absence. She is given brief mention in letters between Philip Rashleigh and his correspondents John Hawkins and Henry Heuland, from which we learn that she specialised in the minerals of northern England, where her husband then lived – hoping to 'exchange his situation against one here [in London] at the Custom House'. In 1807, based at 14 Frith Street, Soho, in London, she was looking for a house in town in which to 'pursue my collecting', but nothing is known of her after that date. She appears to have been a knowledgeable mineralogist and to have had good contacts at the northern mines, which she toured herself from time to time in search of specimens. In 1789 Rashleigh informed Hawkins of her recently acquired 'Collection of Spars etc. from the north of England', in consequence of which Hawkins visited her at Leicester House, London and bought:

> ... *several fine specimens of Fluors & calcareous spars. ... Before the receipt of your letter I had not heard word of this sale so that the best specimens had already been picked. I thought the prices not unreasonable and it gave me pleasure to lay out any money with so intelligent and genteel a female mineralogist from whom I shall expect every year an accession of interesting specimens.*

She seems to have been discriminating in her stock:

> *I have just returned from the different mining counties that produce Spars and have not been able to procure more than 20 specimens that were worth bringing home.*

The specimens mentioned in the Philip Rashleigh MS catalogue as having been obtained from her are nearly all from the Cumberland and Durham lead mines.

William Nicol FRS (1768-1851)

Known as a mineralogist and inventor of the Nicol prism, Nicol is famous for devising a method of making thin slices of fossils, rocks and minerals for microscopic examination. His mineral collection certainly contained northern England material, and a specimen of alstonite (BM 41707) is illustrated. His collection of minerals was eventually acquired by Alexander Bryson.

Revd Clayton Mourdant Cracherode FRS (1730-99)

Although ordained a minister of the Church of England, Cracherode devoted his life to the collection of books, art and nature. His collections were bequeathed to the nation in 1799. Of these, the shell and mineral collections went to the Natural History Museum, London. The catalogue of the Cracherode mineral collection is extant. Cracherode had a comfortable, maybe ample, income; his intention was to build a library and collections of international importance. He was a Trustee of the British Museum. The catalogue of Cracherode's minerals was made by Mr George Humphrey, dealer in shells and minerals, who appears to have been the major specimen supplier to Cracherode. The catalogue lists the prices paid.

John P. Walton (1838-1915)

Walton resided at Akeham High House, Hexham, Northumberland. His collection is known to have contained fine specimens of fluorite, smithsonite, alstonite and witherite.

Sir Arthur Russell (1878-1964)

Russell assembled a magnificent suite of minerals from the north of England in the early part of the 20th century. The Russell collection is now in the Natural History Museum, London. It is particularly rich in contemporary material from the northern Pennines, and specimens from the west Cumbria iron orefield, the latter mostly acquired by the purchase of old collections. Russell was especially proud of his remarkable collection of variously coloured northern England fluorites, mostly from the northern Pennine orefield. A good proportion of the north Pennine material was self-collected. The collection is therefore rich in material from Nenthead, Allenheads, Boltsburn, Nentsberry and Settlingstones. Russell was very active in travelling to the mines working in his lifetime, and establishing a network of contacts with the mine managers, miners and private collectors in the area. For instance, from his diaries we know that he visited the Boltsburn mine and went underground with Mr Willis, the mine man-

SIR ARTHUR RUSSELL

Sir Arthur Russell (1878-1964), perhaps the 20th century's most significant collector of northern England minerals. (©NHMPL)

ager. He records that the workings in the Great Limestone were yielding a marvellous profusion of galena, fluorite, calcite and siderite specimens; epimorphs after fluorite were also collected. We can gauge the efficiency of his contact network from his correspondence, now archived at the Natural History Museum, London. Certainly he was kept well aware of new finds in the working mines. The following two extracts from Russell correspondence gives the essence of the nature of information received, and his keenness to obtain specimens from mines and collectors:

Dear Mr. Russell,
Mr. Harrison and I went to Nentsberry on January 22nd. I have not been able to collect very much in the way of specimens. We are working on the heading in the First Sun Vein just East of the

Liverick vein. I had a walk forward one day and found Witherite crystals on the side, just small ones but very perfect. Also found witherite growing in small round balls. This is all the specimen hunting I have been able to do. Two foremen per shift plus fifteen men. Not much opportunity for hunting specimens. Mr. P. Blight and I were in the Ashgill Force mine a week last Saturday; but did not see any good specimens, very little of any kind.

Mr. Harrison and I still to go to Rodderup on alternate Saturdays to accompany Mr. Martindale in the mine and do odd jobs. We shall expect to see you when you come into the district again.

Yours faithfully,
H Millican

Dear Mr. Russell
I was pleased to have your letter of the 29th. Glad you got home safe and enjoyed your visit to bonny Northumberland.

I got the sample from Donaldsons window sill, at Allendale, he brought all the pieces he had on the window but this is the only one worth sending so it must be the piece you want.

The men will not miss spar you need not have any fear about that, but as I said you had arrived in the nick of time none being seen for quite a long while.

I shall try and find you some samples from the Allenheads mine and any I can find to our mutual satisfaction.

I shall keep you informed if I get any.
Now I must finish with kind regards.

Yours faithfully
J. A. R. Graham

N.B. The East End is stopped altogether no more spar, the men are gone out of it.

As well as Millican and Graham, other important Russell contacts in the north Pennines and Lake District were:

— Lancelot Laverick (1930, foreman Nentsberry mine)

- J. Robinson (1932, foreman of mine)
- John Angus (1934, foreman of mine)
- Percy Blight (1933, Boltsburn)
- Joseph Heller (1923, Coniston miner) who sold to Russell, chloanthite, rammelsbergite and skutterudite from Coniston

Russell visited Fallowfield mine at Hexham, with J. P. Walton the manager, in 1930.

Russell also had a professional interest in the mines of the northern Pennines as, for a time during World War I, he was seconded as a Lieutenant, Royal Engineers, to write a report on the mineralisation of the New Brancepeth colliery in Co. Durham. Russell's report was entitled *The New Brancepeth Colliery (Barytes section) Durham.* Messrs Cochrane and Co. were the owners of the mine and were proposing to extend their milling facility in order to increase the output of ground barite. At this site barite and smaller amounts of witherite were found occupying an important fissure vein. The ground barite was used in both the paint and chemical industries. In this report Russell also gave a description of the dressing plant.

As much of the witherite as possible is picked out from the barytes at the working faces. The two minerals are then conveyed along the main colliery haulage road in ordinary coal tubs to the shaft. At the surface the contents are tipped into 1 1/2 ton wagons and taken to the barytes mill nearby. The barytes is first put on a screen with 1/2 inch holes, and the smalls removed, the large lumps then go to a revolving washer and from thence to a picking belt where the various grades are then hand picked. The rough lumps from the belt are then crushed and delivered by elevators and conveyor along with the smalls from the first screen to a Luhrig washer and vanners. The product consisting of pure barytes is then dried in brick bee-hive shaped ovens (13 in number) each oven having a maximum capacity of 4 tons. The drying occupies about 10 hours, after which the material is ground in ball mills (2 Krup and 1 Edgar Allen being employed), and elevated to French buhr mills 4 feet 6 inches in diameter; it passes through three of these and from thence to a series of dressers, 160 mesh to an inch,

the final product being then bagged in 2 cwt sacks. The finished product is classified in seven grades, according to colour; these grades are (1) Lily White, (2) Best White, (3) No. 1 White, (4) W.W., (5) A.C., (6) A., (7) 3.

The whole output of witherite, which amounts to about five tons per week, is converted at the mine, by treatment with HCl to barium chloride.

The capacity of the existing mill is 150 tons of finished product per week, the mill running day and night and Sundays. The output per annum is roughly 7,000 tons.

Henry Ludlam (1822-80)

Ludlam was a hosier and mineral collector of 174 Piccadilly, London, whose collection was bequeathed to the Museum of Practical Geology. It is now kept as an individual collection in the Natural History Museum, London. Ludlam amassed a very important collection of British Minerals (incorporating the Nevill collection and Forster-Heuland collection), mostly small specimens of classic north of England species.

R. P. Greg (1826-1906) and W. G. Lettsom (1805-87)

Greg was a businessman and Lettsom a career diplomat and collector whose collection was bought by Nevill, and sold to Ludlam in 1877. Greg and Lettsom's *Manual of the Mineralogy of Great Britain and Ireland* was published in 1858 and re-published in 1977. The Allan-Greg collection contained important early north of England material. Alstonite specimens in this collection are all labelled 'Bromley Hill'.

Revd J. J. Couzens (d.1921)

Residing at 3 Elm Grove, Taunton, Couzens did not always show himself as 'Revd' in the Taunton directories, but he was a Congregational minister and a keen collector of fossils and minerals. His mineral collection was mostly of west Cumbrian iron orefield material, and specimens were mainly acquired by purchase from John Graves. Specimens in the collection were of very good quality, especially many fine calcite twins.

Characteristically, Couzens specimens have attached small labels extolling the quality, such as 'Best ever found', 'Finest seen', *etc*. This could be a reflection of Graves' description for the specimen. He became acquainted with Charles Trelease when the latter was living in Taunton. In 1921 Couzens died and his minerals and fossils were purchased by Charles Trelease (1857-1927) of Cornwall and latterly Bristol, and it was from Mrs Trelease, on the death of Charles Trelease, that Russell purchased the collection.

Colonel John Wilson Rimington (1832-1909)

Based in Norlake, Surrey, Rimington amassed a very large collection during the 19th century which contained important north of England material including some remarkable calcite twins from west Cumbria. Eventually his holdings were auctioned in London in 1891 and 1892, and Paris in 1912. Two further auctions were held in Paris after his death, at which time the Paris museums obtained many good specimens. Some fine calcite twins are in the Russell collection at the Natural History Museum, and other specimens form part of the Reading University collection. The Rimington collection had very characteristic labels with printed scroll margins.

John Ruskin (1819-1900)

Ruskin, a poet, artist, philosopher and critic, was one of the great figures of the Victorian age. However, it has been recorded that one of his earliest ambitions included the writing of a mineralogical dictionary. It was his interest in mineralogy that so delighted him when he acquired his first box of minerals. This continuing interest led him to become very well acquainted with, and a student of, the mineral collections of the Natural History Museum (then the British Museum), to which he returned at intervals throughout his life. He even helped Dr Lazarus Fletcher (Keeper of Mineralogy) with the artistic arrangement of the displays in the newly-created Mineral Gallery at the Museum. The collection was moved in 1880 to South Kensington.

There is no doubt that Ruskin was an avid collector of minerals, acquiring specimens on geological expeditions with his manservant, or by purchasing from the important dealers of the day, such as Tennant, Talling and the Bryce Wrights. It is believed that he had a collection of some 3000 specimens. In the latter part of his life Ruskin moved to Brantwood, near Coniston in the Lake District. From the evidence of Ruskin specimens in the Natural History Museum and elsewhere, his mineral collection contained specimens from north of England localities, although there is doubt as to

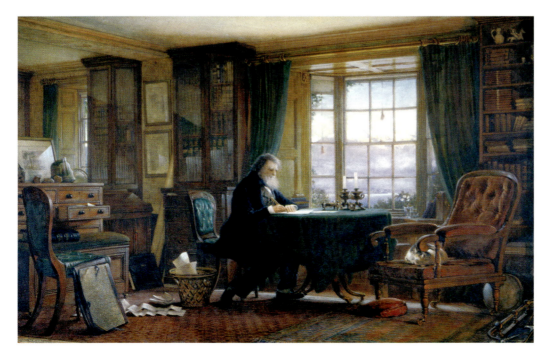

JOHN RUSKIN

Watercolour by W. G. Collingwood (1854-1932)

John Ruskin in his study at Brantwood, 1881. A few of Ruskin's mineral specimens may be seen on the cabinet in the left of the picture. (The Ruskin Museum, Coniston, Cumbria)

whether the specimen labels are by Ruskin. He was a great believer in education. 'Ethics of the Dust' was a series of ten lectures given by Ruskin to 'little house-wives' on the elements of crystallisation. He thought it was most important that mineralogy should be a subject taught in schools, and to popularise these ideas he gave mineralogical talks and donated parts of his collection to a number of educational establishments such as Somerville and Balliol Colleges, Oxford, David School, Reigate and Harrow School. Not all of Ruskin's collection specimens can be recognised with any certainty and much of the original material has been lost. However, the best of the remaining collections are well displayed in Sheffield, in the Ruskin Gallery.

Some Ruskin specimens and mineralogical ephemera are incorporated into the displays of the Ruskin Museum at Coniston in the Lake District. Part of his collection was kept at Brantwood and this was later acquired by Mr Arthur Severn. It mostly consisted of specimens purchased from R. Talling.

John Postlethwaite (1840-1925)

Postlethwaite was born at Birkbeck, in the Vale of New-lands, at a time when this area was busy with mining activity. Mining was in the family, for his father was the under-manager at Force Crag mine, Coledale, and later a mining engineer at the Brandelhow mines. When of age he was to work underground at this mine. Eventually the mines declined and he became a booking clerk, accountant and finally head of the accounts department for the Cockermouth, Keswick and Penrith railway. Despite these office-based jobs, his interest in geology continued and he was to make important Lake District fossil and mineral collections. In 1877 he published the first edition of *Mines and Mining*, and in 1897 *The Geology of the English Lake District with notes on the Minerals*.

Today Postlethwaite's collections are in the collections of Tullie House Museum, Carlisle.

J G Goodchild (1844-1906)

Goodchild was a contemporary of Postlethwaite and collaborated with him on a description of Skiddaw Slate fossil fauna published in 1886. He was a profes-

sional geologist working for the Geological Survey of Great Britain. Some specimens went to his son W. Goodchild, and then to the Bob King collection. Other specimens are today in the collections of Tullie House Museum, Carlisle.

Recent Collectors

There are still many fine collections of northern England minerals in private hands, but detail is only given here of a handful of important contemporary collectors, whose material is on public view.

Dr R. J. (Bob) King

King retired from his post as principal curator in the Department of Geology, University of Leicester, in 1983. Both his hobby and research interests continue to be in the topographic mineralogy of the British Isles. His numerous publications and collections reflect this interest. In 1985 the then Museum of Wales acquired the R. J. King Mineral collection, one of the most important modern private mineral collections, containing a fine representation of north England material especially strong in fluorite from Weardale and Cumbria. A catalogue of the collection, which is now housed in the National Museum Wales, in Cardiff, is in preparation.

Ralph Sutcliffe

A businessman from Lancashire, Sutcliffe twice amassed important collections of northern England minerals. The first, started in the 1960s, was sold into the USA, and the second was sold to Lindsay and Pat Greenbank of Kendal, Cumbria.

Michael and Brenda Sutcliffe with Lindsay and Pat Greenbank

The Sutcliffes and Greenbanks formed a partnership in the early 1970s, known as the Cumbria Mining and Mineral Co., to mine specimens on a commercial basis. In Weardale permission was obtained to work the mineralisation in Rogerley Quarry, Frosterley, Weardale. A small flat has been worked there intermittently over succeeding years and has provided many fine green fluorite specimens. This small mine was acquired in May 1997 by a partnership of American dealers including Jesse Fisher, and was worked on a rather expanded scale under the name of UK Mining Ventures. Worked only for an abundance of very fine green fluorite specimens, it has been the only such venture known to the authors in the British Isles and is therefore unique in British mineral collecting history.

Force Crag mine at Braithwaite, Cumbria has a long history of working lead and barite since the 1860s. The mine has been re-opened several times for industrial barite mining. New Coledale Mining Ltd, with Lindsay Greenbank as a director, began working the deposit in the early 1980s. Many good specimens of barite, sphalerite, galena and siderite were mined. After early indications that commercial extraction of zinc and barite might succeed, the lowest entrance to the mine collapsed and the mine flooded. Abandoned in 1991, Force Crag mine was the last working mine in the Lake District. The National Trust, owners of the mine, have conserved the remaining mine buildings and machinery.

Specimens collected by, or supplied by the Greenbanks are to be seen in several museums in the north of England.

Elizabeth and David Hacker

The Hackers assembled an impressive collection of northern Pennines minerals during the later years of the 20th century, including some notable fluorite specimens. A large number of these were acquired by Killhope Lead Mining Museum in Weardale in 2006 where they form a striking part of the permanent display of local minerals.

Observers

Many famous artists and poets have strong connections with the north of England, particularly the Lake District. Their mineralogical and mining connections may be seen as somewhat tenuous, but their observations are interesting.

John Constable (1776-1837)

Constable was a famous English landscape artist. For most of his working life, pencil was the tool Constable most favoured for his sketches from nature. The pencils that were available in Constable's period consisted of graphite, which was being mined in the Seathwaite area of Borrowdale in the Lake District. This deposit had been worked since the middle of the 16th century. Graphite as a sketching medium had succeeded the use of early metallic drawing instruments (metal-point – plumbago). Due to its similarity to lead, we arrive at the erroneously named 'lead pencil' of today. Constable was drawing with graphite in Borrowdale in 1806, only a mile or two from the source of the graphite. The first graphite (locally termed 'wadd') found at Borrowdale was, for three centuries, the most prized European deposit, and the position of the exact location was jealously guarded. At this time English pencils were made from thin strips of natural graphite set into hollow cylinders of wood. The famous Cumberland Pencil Company began making fine art pencils in Keswick in 1832, although the closure of the Keswick factory and removal of pencil-making to a new site near Workington was announced in 2007.

Joseph Mallord Turner (1779-1851)

In 1797 Turner undertook a famous tour of northern England sketching local scenes and especially views in the Lake District. However, only one sketch of this large collection depicts a mining scene, that is *The lead mine at Grasmere*. However, the view is typically Turner and the outline of detail and information is rather vague. Turner also produced a well-known watercolour of High Force, in Teesdale.

William Wordsworth (1770-1850)

Wordsworth was perhaps the most famous literary inhabitant of the Lake District. He is usually considered to have had little interest in science and indeed had directly castigated geologists in the 'Excursion', published in 1814. However, studies by John Wyatt suggest that there is evidence that Wordsworth was widely read in natural history early in his life, particularly from travel accounts, and developed an analytical eye for the landscape. In due course Wordsworth developed friendships with several senior British geologists such as Adam Sedgwick. It is doubtful that Wordsworth at any time had a collection of geological material of Lake District minerals and rocks, but Wyatt considers that Wordsworth's later poems demonstrate themes that raise the study of geology to one of high purpose.

L. S. Lowry (1887-1976)

The artist is best known for his paintings of 'matchstick' figures and Lancashire mills. However, a drawing by him in pencil and ballpoint came to light in an exhibition at Sunderland Museum in 1987. The drawing, labelled (not in the artist's hand) 'Old mill, Whitfield Moor, near Alston', was immediately recognised as the Killhope Wheel (now the Killhope Lead Mining Museum) and associated buildings as they were in 1963, the date on the picture.

W. H. Auden (1907-73)

The poet was particularly fond of the northern Pennines, with the village of Rookhope, Weardale, a special centre of affection. It is known that in later life he was a fan of railways, leading to perhaps his most well-known poem 'Night Mail', published in 1935.

His interest in things of an industrial and archaeological nature probably led from a young age to his interest in the old lead mines of Weardale. One poem, 'The Old Mine', appears to be based on his recollection of a lead mine at Rookhope, Weardale, visited while on holiday as a child. This 1924 poem is partially based on the mine and the Rookhope to Bolts Law inclined railway. His family was at this time living at Threlkeld, in the northern Lake District.

Private Museums

Another remarkable feature, falling somewhere between collections and dealers, was the early development of at least three separate private museums in the Lake District. It is doubtful whether any of the mineral specimens displayed in these establishments can now be recognised, but there is no doubting that their contribution to the general knowledge of northern England mineralogy was significant.

PRIVATE MUSEUM ADVERTISEMENTS

Advertisements for Hutton and Shelton's, and Crosthwaites private museums, Keswick. (©NHMPL)

Crosthwaite's Museum (1780-1870)
Keswick, Cumberland

Crosthwaite's Museum was founded in 1780 by Peter Crosthwaite (1735-1808), a native of Dale Head, Thirlmere, who had spent seven years abroad in the East India Company and twelve years as a Customs Officer in Blyth, Northumberland, before settling in Keswick in 1780. He opened his collection to the public in 1781 to such success that he built the substantial Museum House in Keswick in 1784 to house his extraordinary cabinet of foreign and local curiosities. Nowadays a plaque in the pavement near the present Herries Thwaite shopping centre commemorates his pioneering enterprise. The museum was known far and wide and contained a remarkably rich variety of objects. The museum and shop were successful enough to provide Crosthwaite's main income, and his careful observations and notes showed him to be passionate about his collecting and collections.

Crosthwaite himself seems to have been one of the most curious items in the museum. This has been said of many a collector and curator, but in Crosthwaite's case there is ample documentary evidence to support the claim. He was, for instance, in the habit of spotting potential customers with the aid of an ingenious set of mirrors, and attracting their attention by means of drums, a barrel-organ, and by the banging of a large Chinese gong, so loud in the open air that it could be heard four miles away on a still day.

His arch rival was fellow Keswick museum owner Thomas Hutton, whose character and actions Crosthwaite vehemently criticised at every opportunity.

Despite this rivalry, and perhaps in part because of it, his museum was popular and the collection excellent. His maps and guides sold in thousands to tourists, and several wrote detailed accounts of their visits. John Ruskin sang its praises in verse. One visitor even remarked that Crosthwaite's daughter Mary was 'an elegant woman and more worth seeing than anything else in his house'. In 1785 Crosthwaite devised the first 'stone xylophone' or 'geological piano' from carefully trimmed and tuned rocks from the bed of the River Greta.

When Peter Crosthwaite died in June 1808, his will bequeathed the museum and business to his son Daniel and daughter Mary.

The fourth floor room served as the museum shop, where mineral specimens were sold. The Crosthwaites offered the following description of the geological collections in 1826:

Two large cupboards, and 16 large drawers, containing the scarcest minerals, and the finest Chrystallisations of Spars, with every variety of the finest colours imaginable. Also 50 varieties of Rock Specimens, all as different as possible, collected … from the most interesting mountains and places in Cumberland. From this collection the company are at liberty to select single specimens, and he sells Cabinets of different descriptions out of the same collection as follows: 100 Superb Specimens … at five pounds, 50 Geological or Rock Specimens … at one pound. … A Lady's small Cabinet of 50 Specimens … for ten shillings.

Thomas Hutton (1754-1831) Museum
Keswick, Cumberland

A museum was established in 1785 in Thomas Hutton's home in direct competition with the successful museum of Peter Crosthwaite. Hutton relied on his established popularity as a tourist guide to make a success of his exhibition, intending to win Crosthwaite's custom and drive him under. Hutton lacked the international experience of Crosthwaite, but through his own efforts and studies had become an able guide with a thorough knowledge of natural history and geology.

Peter Crosthwaite was furious at the interloper, and criticised him vehemently in public and in his journals. However, Hutton had his supporters as a guide, described by one of his clients:

Hutton, our guide, and keeper of a second-rate Museum, is a very intelligent man; he is very clever as a Mineralogist and Botanist, but not so absorbed in science as to prevent his being a most entertaining travelling companion; he related many anecdotes and his descriptions of the scenery are correct and sufficiently diffuse, without being dry or lectitious. He is a grand-father, and of plain, farmer-like appearance.

Hutton's museum was near the Royal Oak Inn, Keswick, and close by he kept pleasure boats. No detailed accounts of the museum and its collections have survived, and it is likely that none was made. It doubtless fulfilled his clients' desires for novelty and curiosity with the usual range of birds, shells, minerals and antiquities, but the variety and quality fell far short of Crosthwaite's collection.

Thomas Hutton died in 1831 aged 85, and in his will he instructed that 'all the Minerals belonging to the Collection of Curiosities, remain as they are' and he bequeathed the museum and his boats to his two daughters, Hannah Hutton and Mary Shelton.

William Todhunter (d.1835) Museum Hawkshead and Kendal

Todhunter's Museum was established in Hawkshead, in the south of the Lake District, in 1784 in a back room of William Todhunter's hat and book shop. In 1793 he moved to Kendal, gateway to the Lakes for visitors from the south, opening first at 177 Highgate in 1796, then moving to the main street at the corner of the Fish Market and Crock Lane four years later. The museum occupied two rooms displaying 'Minerals, Shells, Petrifactions, Incrustations, Crystallisations, Spars, Marbles, and many curious Fossils, Mosses, Lichens and plants of spontaneous growth, and a variety of Birds, Quadrupeds, Fishes, and Coins, Medals, Antiques, and curiosities'. Like many other museum collections of the time, documentation was poor and we have little idea of the full extent or importance of the museum. From it he sold the usual variety of minerals, spars and petrifactions. He supplied minerals and musical stones to James Sowerby 1806-25 and must have stocked more than just tourist pieces.

Todhunter died sometime before July 1835 when the museum of '... the late Mr Todhunter' was offered for sale at auction including its 'Large fine Chrystallisations; rare undescribed Fossils, and others; Ores and singular Combinations, &c, some Foreign Minerals and Precious Stones; sets of Minerals and Musical Stones ...'.

In the following August a new museum came into being with the newly-formed Kendal Natural History and Scientific Society into which were incorporated some of Todhunter's exhibits. For some decades the society and its collections flourished, especially under the presidency of the great geologist Adam Sedgwick (1796-1876). The society declined towards the end of the century and in 1910 parts of the collection were sold to the British Museum (Natural History), now the Natural History Museum, London, and the remainder passed to the local council to form the basis of a municipal museum.

Mineral Dealers

Whilst the mines of the north of England were active – indeed booming – throughout the area, there were professional mineral dealers. The earliest were probably centred on the Alston area during the 19th century. Few of these dealers were full time and most had other trades and dealt in minerals as a sideline. There is little doubt that the most important dealers in northern England minerals were John Graves of Frizington, the Gilmores of Alston, and Bryce McMurdo Wright of Caldbeck. Between them they supplied many fine specimens to collections worldwide. In fact, judging by the extensive advertising of northern England specimens by many internationally important American dealers in various volumes of the *Mineral Collector*, the heyday of international trade was in the late 1890s.

Brief comments follow on a selection of the more important dealers.

As already stated, many of the early British mineralogical texts contain catalogue entries for specimens from the north of England, although locality details for these early specimens are often not of the highest standard. For instance, many specimens from the Alston area were simply shown as from Alston Moor, a 'bucket' term for a large number of individual mines and workings. Perhaps the lack of detail on locality information was deliberate, the dealer (or collector) letting the specimens speak for themselves and protecting their sources, whether it be locality or miner! This lack of accuracy in locality detail could often be misleading: for instance, early specimens of alstonite were known as 'bromlite', because the early locality given for their source was in error as Bromley Hill, rather than the true designation as Brownley Hill mine.

The Gilmores of Alston

The Gilmores were active dealers in the late 1800s. In the collections of the Natural History Museum, London, many specimens of Fallowfield witherite were purchased from them between 1881-92. One of these witherites is figured in *Miers Mineralogy* (1929). There seem to have been three active 'mineral dealer' members of the Gilmore family. Patrick Gilmore (father) (1833-92), Peter Gilmore (son) (1856-92), and Patrick's wife, Elizabeth Gilmore, who continued the business for some time after the demise of father and son. (The year 1892 was not a good one for the Gilmores; the father Patrick collapsed and died at his son's funeral.)

An advert in the Alston directory describes Patrick Gilmore as a wholesale mineral dealer. In the 1884 directory the trade of Patrick Gilmore is shown as a marine-store dealer and mineral dealer. We know that Gilmore was visited by the American collector Charles Pennypacker in 1889. Pennypacker records that he spent several days with Patrick Gilmore whom he described as a junk dealer and with a warehouse full of specimens. In 1896 Elizabeth Gilmore moved to Barrack Street, Carlisle, where she continued the business for some time. In 1897 a letter offering galena also shows 'the desire to sell many pieces as I find much to do with my family, and there has not been such demand for minerals lately'.

John Cowper senior, John Cowper junior and family

Based in Alston, the Cowpers were important dealers during the 1800s. They also had a shop in Keswick. John senior appears to have been an active dealer between 1828-51, being shown in various trade directories during this period as a 'mineralist and jeweller'. It is known that the Cowper family exhibited barite, witherite and other north of England minerals at the *Great Exhibition* in the Crystal Palace, Hyde Park, London, 1851. The following recommendation for Cowper was given by Mantell in 1844:

A complete series of the minerals of Cumberland, comprising specimens of great beauty and interest, can be obtained of John Cowper, Alston, Cumberland. This collector formed one of the finest series I have ever seen for the Revd Charles Pritchard, FRS: he may be relied upon for his knowledge and attention, and his prices are moderate.

'I have long been an extensive dealer in minerals at Alston Moor,' wrote John Cowper senior by way of introduction to Professor E. D. Clarke of Cambridge University in 1819 (Clarke Archive). From this, and from fragmentary evidence, it appears that the Cowpers were considered important and successful local mineral dealers. They did not make a living from minerals alone; jewellery and groceries made their contribution and, on one occasion, unable to make an arranged meeting with Clarke, John sent 'a brace of Grouse and a few minerals, which I beg your acceptance of'.

By 1851 John Cowper senior was dead. Sarah, his wife, described herself as a mineralogist in the 1851 census, although she was by then 71 years old. Only her sons appear in subsequent trade directories.

William Burrow (1791-1877)

Based in Alston, Burrow appears to have been a significant dealer in minerals. Little is known of the man other than as a name in trade directories, where he is listed as 'Burrow', or 'Borrow', in the period 1828-77, as a mineral dealer. He seems to have been a full-time professional dealer for over 50 years, during the most productive years of the northern England mining fields. He supplied many specimens to the British Museum (Natural History), now the Natural History Museum, and to other collections. Material supplied by Burrow was usually of good quality.

John Birkett (1836-97)

Birkett was a mineral dealer in Keswick. Specimens supplied by him can be found in many museum collections. He was described as a watchmaker, jeweller, gunmaker and mineral dealer of Market Place, Keswick. When the Keswick Literary and Natural History Society founded a museum in the Moot Hall in 1875,

Birkett became one of the honorary curators, along with the well-known geologist Clifton Ward. William Greenip took charge of the museum in 1883, Birkett remaining as curator, a post he held until at least 1889. Birkett continued to appear as a mineral dealer, jeweller, etc., in trade directories until 1897.

John Graves (1842-1928)

John Graves of Frizington was one of the most influential and important north of England dealers. He was most active in the trading of fine mineral specimens from the west Cumberland iron orefield and Alston Moor, and supplied many extraordinarily fine specimens to museums and institutions worldwide. Volume 1 of the *Mineral Collector* (1895) carries an advert for his dealership. Many 'Graves' specimens crossed the Atlantic at the turn of the century as stock for some of the famous American dealers of the time. These dealers widely advertised the arrival of fine new material from northern England. Amongst them were Dr A. E. Foote of Philadelphia, George English of New York, Lazard Kahn of New York, and Charles Pennypacker. The American dealer George English eventually purchased most of the business stock and private collection in about 1920, just before Graves' death. Some specimens

remained in the family and were eventually purchased by the London dealer P. Botley in 1942-43.

John Graves was born in Langdale, Westmorland. It is said that as a boy in the 1840s he moved to the village of Frizington, west Cumberland, the very centre of the iron ore mining field. It appears that Graves soon became interested in the sparkling crystals he found on some of the mine dumps. It is further suggested that Graves became a miner in order to recover some of the fine calcite, quartz, barite, dolomite and fluorite crystal groups. By his own reckoning he started collecting seriously in 1873, but he was still shown as an iron miner in the 1881 census. So too was his eldest son George, who was then 14. Perhaps he turned profes-

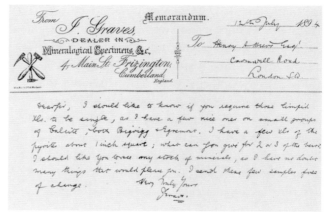

ABOVE: JOHN GRAVES' LETTER

Letter from John Graves of Frizington to Henry Miers, Keeper of Mineralogy at the British Museum (Natural History), now the Natural History Museum, 12 July 1894. (©NHMPL)

RIGHT: DEALERS' ADVERTISEMENTS

RIGHT: John Birkett of Keswick, advertisement late 19th century. (©NHMPL)

FAR RIGHT: John Graves of Frizington, advertisement in the *Mineral Collector*, 1895. (©NHMPL)

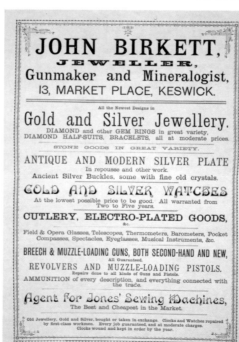

sional dealer in 1882. His collection and sources of material obviously became large enough for him to operate full time as a professional dealer. He lived and operated his business from 4 Main Street, Frizington, and although his style of business letter-head changed often, this is where he lived for the rest of his life. As well as his international trade he of course provided specimens to most of the major British museums. His characteristic trade correspondence and invoices are to be found in archives of these institutions. The essence of most of his handwritten letters to such as Lazarus Fletcher and Spencer (keepers of mineralogy at the British Museum [Natural History]) and Miers (at the Oxford Museum), always extolled the rarity, beauty and perfection of his trading specimens and the fact that the specimen offered was probably the best ever raised. He always invited clients to visit his collection in Frizington, probably a sensible move considering problems of freight carriage, and so spread the word of his business. Both specimens for the home market and overseas had to be carefully packed and were usually sent by rail, although some breakages were reported. After the turn of the century many of his letters implied that he was not in good health and due to pecuniary circumstances he asked for interest or names of collectors who would be prepared to purchase the bulk of his stock.

In 1925 Graves suffered a stroke which left him partially paralysed for some time. He died in 1928 and was buried in St Paul's churchyard, Frizington. His son, F. W. Graves, carried on the business for a while after his father's death, but mineral collecting was becoming moribund in the British Isles reflecting the general decline in metalliferous mining. Graves' testimonial is the presence of his specimens in major mineral displays worldwide.

Charles H. Wright (d.1865)

Wright was a well-known Keswick guide and notorious storyteller, who accompanied men such as Adam Sedgwick and John Ruskin on tours of the Lake District. In a letter to Revd Peter B. Brodie on the 10th of September 1854, Sedgwick advised him to

... find out Charles Wright – a guide formerly; and now, I am told a guide director. You must take what he says cum grano salis, *for he is a bouncer.*

Wright served as a guide to the Ruskins in their tour of the Lake District in 1837 and John Ruskin later commented:

Wright at Keswick knows more about the country than any other guide; but don't believe anything he tells you about anything but rocks.

In addition to being a 'scientific guide' and dealer in minerals, Wright also advertised himself as a 'bird and animal preserver', botanist and geologist, and ran a lodging house in St John Street from at least 1858. Sometime around or before 1869 the business was taken over by Maria Wright, possibly his wife or, from an entry as 'Miss Wright' in the 1882 directory, his daughter, who featured variously as botanist, mineral dealer and lodging-house keeper in directories up to 1884.

Arthur Hutchinson (1819-85)

Based in Keswick, Cumberland, Hutchinson was listed among 'dealers in minerals, precious stones, shells' in Bulmer's 1883 Cumberland directory. He sold specimens to the British Museum, but as often as not failed to meet the museum's quality criteria or tight budget.

... I know I have the best and most valuable specimens in the north of England, I travel and collect them myself from all the mining districts in the North, I sent 5 specimens up to the British Museum on the 4th day of April 1867, 1 large Calcite among them I got £5.0.0 for it.

Bryce McMurdo Wright (1814-75)

Wright was a very influential dealer in minerals of northern England. He was born in Dumfries, but liked to be thought of as a native of Caldbeck having lived in Hesket Newmarket, Caldbeck for many years, and having once been a miner in Cumberland. He became a mineral dealer in Liverpool (1843-55) and

London (1855-75), becoming the most successful British dealer in minerals, fossils and shells in the mid-19th century. He supplied north of England specimens to many important collectors, international collections and dealers. When he was in the field his wife Jane, and later his daughter Elizabeth and son Bryce McMurdo junior, ran the shop. The business continued under his son until the latter's untimely death in 1895. Bryce Wright can now be recognised as something of a specialist in north of England minerals, especially those related to some of his Caldbeck Fells specialities, like Roughton Gill hemimorphite and pyromorphite.

William Greenip and Son

Based in Keswick, Cumberland, the Greenips were active from 1829 to 1880, and in business from at least 1855. Greenip senior was described variously in directories as naturalist, dealer in minerals, and entomologist. Postlethwaite noted in 1913 that Greenip had a stone xylophone that he had made in the 1870s. Both served as local postmen in Newlands and Braithwaite. Greenip junior took charge of the museum of the Keswick Literary and Natural History Society in 1883.

William (Bill) F. Davidson (1907-2002)

William Davidson of Penrith was a dealer of some importance during the 1950s and '60s. During the Second World War he was a special constable, and was able to obtain minerals from several of the then working mines. He is best known for supplying barite from Crowgarth mine, blue fluorite from Ullcoats mine, calcite from Rodderup Fell mine, and barite from Cow Green and Silverband mines. He provided many fine specimens to Scott J. Williams the American dealer, of Minerals Unlimited, Scottsdale, Arizona. Scott Williams purchased Davidson's private collection in the 1950s.

Joseph Graham

Joseph Graham of Keswick, Cumberland was active as a dealer between 1829-69 and was described by Bolton in 1869 as 'poor old Joseph Graham, a mineral dealer and shoemaker, in Keswick'.

Anthony Furnace (1823-98)

Furnace of Keswick, Cumberland was described as a jeweller, watchmaker and mineral dealer, and is known to have been active between 1858-97.

John Pattinson

Based in Alston, Pattinson is known to have been in business in 1829 as a mineral dealer as well as a printer and bookseller.

Collecting and Heritage

Many of the specimens from the private collections of the past have now found their way into national and local museums where they are available for study.

Pre-eminent amongst collections in the United Kingdom is that in the Natural History Museum, London (formerly known as the British Museum [Natural History]). Most of the specimens illustrated in this book are from this collection. Of special note within this museum is the large collection, assembled during his long life by Sir Arthur Russell, and bequeathed to the museum in 1964. Notable suites of specimens in the Russell collection include spectacular examples of northern Pennine fluorite and witherite and west Cumbrian calcite.

National Museums Scotland holds important specimens of northern England minerals, several of which are also illustrated in this book.

National Museum Wales houses the Bob King collection in which are numerous striking specimens of minerals from northern England.

Other important collections of northern England material can be seen and studied in several regional museums throughout the United Kingdom. Notable among these are the following collections:

- World Museum Liverpool – which has a fine suite of old and contemporary northern England specimens, including a very spectacular large example of purple fluorite mined during the 1980s at Frazer's Hush mine, Weardale.
- Manchester University Museum.

- Tullie House Museum, Carlisle – which contains Postlethwaite and Goodchild collection material.
- Cleveland Museum of Natural History, Middlesbrough.
- Yorkshire Museum, York.
- Hancock Museum, Newcastle-upon-Tyne.
- Sunderland Museum.

A smaller, but extremely fine collection of northern Pennine minerals, including some striking fluorite collected by Elizabeth and David Hacker, is to be seen at Killhope Lead Mining Museum, Weardale. This museum also holds a unique and comprehensive collection of spar boxes from both the northern Pennines and west Cumbria.

Worldwide, virtually all of the major museums contain a remarkable number of northern England calcite, barite, fluorite and witherite specimens. Clearly the early dealers did their job exceedingly well – trading and dispersing specimens around the world. For example, the magnificent collection of west Cumbrian barite in Freiberg Academy, Germany, deserves the attention of all collectors and students of northern England minerals.

Individual Heritage Sites

Many years of abandonment and neglect, together with numerous land reclamation projects across the region, have obliterated the remains of many mining sites, although spoil heaps and mine buildings can still be seen locally. A wealth of books and leaflets describe the mining and related social history of many of the region's former mining areas.

Many of the wild and desolate landscapes of northern England are vividly described in the books by Alfred Wainwright (1907-91). These have inspired walkers and visitors to the fells and dales and his publications contain many references to the remains of the mining industry. His seven guide books to the Lakeland Fells were written between 1952 and 1966. In 1973 he published *A Coast to Coast Walk from St Bees to Robin Hood's Bay*.

The designation, in 2003, of the north Pennines Area of Outstanding Natural Beauty (AONB) as a Global Geopark recognised the important relationships which exist in this area between the geology and mineral deposits and the physical, cultural and economic landscape. Exciting activity in the conservation and interpretation of metalliferous mining heritage sites of northern England, in part encouraged by this prestigious designation, has occurred in recent years and it is hoped that further developments will flourish in the coming years.

Of these the mine and mill site at Killhope Lead Mining Museum is one of the best mining interpretation sites in Britain. It consists of a full reconstruction of the surface workings of a typical 19th-century north Pennines lead mine, including the famous water wheel. The Park Level adit has been re-opened as an underground experience to provide access to a replica of a

BILL FROM JOHN GRAVES

Sent to Henry Miers. at that time at Oxford University Museum (26 May 1899). (©NHMPL)

typical northern Pennines lead mine. The miners' lodging house, museum and café make the site a must for both education and tourism.

Nenthead Mines Visitor Centre, operated by the North Pennines Heritage Trust, contains displays of the local geology, natural history and the history of lead and zinc mining in and around the village. Underground tours are available in part of Carr's Level mine, adjacent to the visitor centre.

At Allenheads, the site of one of the largest lead mines in the area, the Armstrong hydraulic engine used in the mine sawmill is preserved in the centre of the village, close to the surface remains of the mine.

The Weardale Museum at St John's Chapel contains important archival documents for the local area, together with a small selection of minerals. In Teesdale, the Bowless Visitor Centre includes a modern interpretation of northern Pennine geology and exhibits relating to local lead and barite mining.

Beamish Museum, County Durham, is an open-air site dramatically displaying life in the North East with working models of collieries and transport systems. The museum also houses important archival collections of objects, documents, and most importantly early mining photographs.

Further south in Yorkshire, the remains of the Grassington Moor mining field is explained by interpretation boards and gives a good introduction to mining developments in the area. The Greenhow mining field, near Pateley Bridge, has much to offer the industrial archaeologist, but as yet there is little interpretation. At Gunnerside the well-preserved Old Gang and Surrender processing and smelt mills can be seen.

Guide books or explanatory leaflets exist for all the above sites for the serious student. In the Lake District, for example, mine trails and interpretation may be seen at the Coniston Copper mines, while the Threlkeld Quarry and Mining Museum and the Keswick Mining Museum provide displays, information and books on the mining history of the area.

Postscript

Although mining has all but ended, responsible collecting and studies by both amateur and professional mineralogists from quarries, spoil heaps and abandoned mines continues to add to our mineralogical knowledge, specimen and mining heritage. Within the last two decades such collecting has revealed two mineral species new to science from within the region: brianyoungite from Brownley Hill mine, Nenthead, and redgillite from Red Gill mine, Caldbeck Fells. These, together with fine specimens of fluorite, alstonite, barytocalcite and witherite from the northern Pennines, and records of numerous rare supergene species from across the region, bear witness to the benefits to mineralogical science that can come from these activities.

KILLHOPE LEAD MINING MUSEUM, WEARDALE

The life and work of a 19th-century lead miner is realistically interpreted at Killhope Lead Mining Museum. As part of the experience, visitors can sample the hard work needed to operate hotching tubs, which enables the separation of heavy galena particles from the lighter waste mineral and rock fragments on the ore-dressing floors. (B. Young)

Introduction
to the Minerals

As outlined earlier in this book, a remarkably large and varied range of minerals is known from northern England. Many are comparatively common species; others are scarce or, in some instances, very rare. In a work of this sort it is not possible to list or to describe all of these. However, the area has long been famous for providing some of the world's finest examples of species such as fluorite, barite and calcite, and includes type localities for no less than five new mineral species: witherite, barytocalcite, alstonite, brianyoungite and redgillite. Spectacular specimens of the first three of these, from northern England localities, are prominent in most of the world's mineral collections. Before looking more closely at a selection of the finest examples of some of the area's minerals, the following pages offer comments on some of the best known of these species.

Calcite

Calcite is, with the exception of quartz, the most common of all minerals. Anywhere that calcareous solutions can penetrate, calcite may be deposited. Like quartz it is sometimes found in remarkably well-formed and brilliant crystals. Calcite crystals have long been the subject of detailed studies by crystallographers, physicists and mineralogists. Calcite's strong double refraction (birefringence) and perfect cleavage attracts the attention of students of mineralogy. Calcite is the most morphologically varied of all mineral species, and to appreciate its vast array of crystal forms and the mechanics of crystal formation is a study for both the amateur and professional mineralogist. Well over 600 different crystallographic forms are known. Count Bournon's treatise on calcite, published in 1808, contained the descriptions of some 56 distinct forms. In Miers' *Mineralogy* (1902) 160 forms are recorded. Charles Palache's literature survey of 1943 documented 630 different calcite forms and Hauy's fascination with calcite is well documented. Whether it is true that he reduced many fine prismatic calcite crystals to dust

OPPOSITE PAGE

Witherite from New Brancepeth Colliery, Co. Durham. Stacked aggregates of white shallow pseudo-hexagonal crystals with some alstonite, barite, sphalerite and coal (max. dimension 15.0 cm). This specimen [BM 1909, 527] was presented by Brodie Cochrane Ltd, of Newcastle-upon-Tyne. Some remarkably fine examples of witherite were found in vein deposits related to the northern Pennines orefield, in a few of the Durham collieries. (©NHMPL)

in an attempt to find a form other than the rhombohedral shape on cleavage fragments is an open question. Certainly these experiments led Hauy to conclude that the external forms of calcite crystals were created by the stacking of minute rhombohedra in various ways. Today sophisticated, quantitative and environmental studies of calcite are aimed at relating the morphology of the crystals to temperature, pressure, pH, the partial pressure of CO_2, solubility, and perhaps the degree of super-saturation. The nature and concentration of impurities and trace elements present is another important consideration. All these studies are aimed at understanding the development of a given form on a growing crystal.

Very fine crystal groups of calcite occur throughout the mines of northern England, especially in some of the lead mines of the north Pennines where calcite, though rarely abundant, is locally a common gangue mineral. However, perhaps some of the most remarkable calcite crystals were those obtained from the west and south Cumbria iron orefields late in the period 1890-1910. In these brilliant crystal groups the crystals are often smooth, bright and transparent. Within the crystal groups, combinations of different forms and crystal habit are important, as is the variety, perfection and size of the twinned crystals. In general terms four distinct habits may be recognised: (1) rhombohedral, (2) prismatic, (3) scalenohedral, and (4) tabular. Combinations of the prism, scalenohedra and rhombohedra form some of the most characteristic crystals of the iron orefields, e.g. from the mines around Bigrigg and Pallaflat. Specimens from this area commonly have both a limonitic and hematitic matrix, but it is hard to be sure of the individual mine source if unlabelled.

These are the crystals found most commonly in museum collections and frequently figured. They are usually water clear but sometimes contain inclusions of reddish hematite (ghosting is common), or black mossy goethite, and in some cases black pyrolusite. Calcite crystals from the Stank mine in the Furness area are usually sharply terminated by acute scalenohedra with little evidence of rhombohedral development. These crystals are often delicately tinted or liberally coated with red hematite.

The heyday of Cumbrian iron ore mining is now long past and so also is the availability of the remarkable crystal groups of calcite, barite, dolomite, etc. However, some mine dumps still exist, and good, but not classic, crystal groups are occasionally collected. The classic crystal groups appear to have been collected at the turn of the last century, but interesting material was collected right up to the almost complete closure of the main mines of the area. For instance, remarkable groups of trigonal, prismatic, calcite crystals were collected in 1979 and 1980 from the Florence and Beckermet mines. Florence miners found the crystal groups in old workings associated with the so called 'banana slide' area between the Beckermet and Haile Moor mines.

Twinning of Calcite

There are several well defined twin laws which form distinctly different groups of twinned calcite crystals. In the famous twin crystals from the west Cumbria orefield, two types of twin with variants are common: (1) scalenohedra twinned on basal pinacoid – such twins are also common in the Derbyshire lead mines and are sometimes known as 'Derbyshire twins'; and (2) the famous 'butterfly' twins. In these, twinning is produced on the face of a rhombohedron.

In such twins the re-entrant angles between the two twin portions are commonly 'filled up' by certain faces, giving different aspects to the crystals and making the crystal morphology of the resultant butterfly twin difficult to interpret.

All these types can be found in the twin calcite crystals of the Egremont area. In 1889 a short but important paper was published in the *Mineralogical Magazine* by Professor H. A. Miers concerning a find in March of that year of calcite twins from the mines in the Egremont area. In this paper he described in some detail the features of the heart-shaped twinned crystals. Many of the individual twins are up to 10 cm in size and examples up to 20 cm are known. It is the large size and the variety of these butterfly twins that gives them international importance. The twin crystals are mostly sparkling and water clear, although sometimes tinged and zoned with disseminated red hematite. Zoned crystals appear to have a 'ghosting' of fine hematite. Today these twins are rarely found in northern England and were rarely found on matrix. Their origin

'PROPELLER TWIN' OF CALCITE, BIGRIGG MINE, NEAR EGREMONT, CUMBRIA

This so-called 'propeller twin' is a most unusual specimen, about 7cm across. It was part of the collection formed by Lady Constance Russell, mother of Sir Arthur Russell, and is now in the Russell collection (reg. no. BM 1964, R 5610) in the Natural History Museum. (©NHMPL)

and preservation within the vugs and their large size is still in some instances unexplained. Most of the larger twins were found in the mines of the Bigrigg area of west Cumbria. Perhaps one of the most remarkable single twin specimens from this orefield is the so-called 'propeller twin'. In this specimen three individual twinned crystals are in a twinned relationship to the central crystal. Other remarkable calcite crystals have been found in other areas of the north of England.

Barite

Barite is a common mineral often associated with metalliferous veins throughout the region, especially those containing lead minerals. It is easily recognised by its weight and cleavage. Its habit is variable but is often tabular; some crystals are elongated along the 'b' axis, and some crystals are rich in forms but without the complexity of calcite.

Many noteworthy localities occur throughout the northern Pennines: for instance at the Dufton Fell and Silverband mines, Cumbria where large clear single

tabular crystals were common. The following extract from the *Mineral Collector* (1902) gives some idea of the sizes obtained and the various uses for barite.

Dufton Fell Mine in Cumberland, England, yielded large single crystals [of barite] *which Bryce M. Wright of London sold. No locality has ever furnished such excellent examples. One crystal weighing thirty pounds is in the collection of William W. Jefferies at West Chester, Pa. For many years it did duty as an anti door-slammer in Mr Jefferies' residence on North High Street. It is about eight to ten inches* [20-25 cm] *high and three inches* [7.5 cm] *or more in diameter, and perfectly terminated.*

Other notable northern Pennine localities for barite include Cow Green mine, Teesdale, from which some fine tabular crystals were collected, and Settlingstones mine, near Hexham, which was long famous for the magnificent groups of sharply terminated chisel-shaped crystals. This distinctive habit seems to be restricted to deposits in which barium carbonate minerals are also common and may be characteristic of barite that has developed by secondary alteration of an original barium carbonate. In Yorkshire fine barite groups, normally white and slightly transparent, and with a morphology reminiscent of the Settlingstones specimens, have over the years been collected from the Wet Groves mine in Wensleydale.

However, it is within the mines of the west Cumberland iron orefield that the north of England's most prized barite specimens were collected. Associated with lustrous calcites, the barite crystals were found in vugs (loughs) within the hematite orebodies. Although there is less variation in form than with calcite, the variety of colour is remarkable, with a range from light blue, green, yellow, amber and brown. Commonly crystal groups have a delicate coating or 'ghosting' of hematite, giving a reddening to the overall body colour of the crystal. Towards the edge of some tabular crystals colour zonation may be important, with the outer zone often a creamy-white. It was some of these colour-zoned crystals that Graves described as barite 'phantoms' and that fetched the highest prices on the mineral specimen market. Virtually every hematite mine produced some barite, but perhaps the two most famous mines for coloured crystals were the Mowbray (light blue-green) and Dalmellington (yellow-amber) mines. Crystals were normally tabular, often with sawtooth edges, but some remarkable 'chisel' like crystals and groups were also found.

In 1930 Sweet showed that some barite from Mowbray mine exhibits striking colour changes when exposed to sunlight. She discovered that several specimens in the collection of the Natural History Museum were pale yellow or amber-yellow when purchased by the museum between 1909 and 1913. However by 1928 these were pale green to sky-blue in colour. To investigate this curious phenomenon, specimens of yellow barite from here, which were known to have been stored in the dark since their collection, were exposed to bright sunshine. Within a little over three hours the original yellow colour had changed to pale green, and after a further two hours it had become distinctly blue-green. No explanation is known to account for this colour change, which appears to be unique to specimens from Mowbray mine.

In the Furness area the Stank mine produced small blue tabular crystals of barite associated with the fine calcite crystals.

In the Lake District, Force Crag mine, near Braithwaite, Cumbria has provided some remarkable groups of white tabular crystals, and good examples of typical 'cock's-comb' form are known from Greenside mine.

Fluorite

Bancroft suggested in his publication 'Gem and Crystal Treasures' (1984) that the demand for English fluorite crystals far exceeds supply. It is certainly true that the north of England has provided a treasure chest of museum quality specimens of colourful and beautifully-formed crystals.

Although fluorite occurs throughout the north of England, the greatest variety of fine coloured fluorite came from the mines and quarries of Weardale. Of all the individual mines, those from the Boltsburn mine, Rookhope are renowned for perfection and variety of colour.

A detailed description was given by Dr Bob King in

his award-winning paper 'The Boltsburn mine' published in the *Mineralogical Record* (1982). Fluorite crystal groups lined many vugs throughout the veins at Boltsburn, but the most spectacular coloured groups originated in the 'flats' within the Great Limestone. Here groups of large simple cubic crystals occurred in cavities within the intensely mineralised limestone adjacent to the vein. Some of the most transparent crystals were hand picked and sold to the Zeiss Optical Co. for the manufacture of apochromatic microscope lenses. Many museums have very large blocks with simple unmodified purple cubes up to 30cm along the cube edge and characteristically fluorescent. Often these cubes have white patches on their faces, probably alteration as a result of a late stage metasomatism. There is a little colour zoning but it is not spectacular. Also collected from these flats, but in lesser amounts, were clear colourless, yellow, amber, brown and olive-green crystals and groups of creamy-white slightly curved opaque cubes.

Equally famous are smaller groups of lustrous purple crystals that average one inch (25mm) across and which exhibit well-developed vicinal structures on their faces.

Modification by the octahedron was uncommon at Boltsburn, but the tetrahedron is occasionally devel-

FLUORITE, FLORENCE MINE, EGREMONT

Clear bright blue cubes on pale cream dolomite from the Ullcoats section of Florence mine. This mine was famous for fine specimens of this unusual colour. The largest crystal is 15mm across. (D. I. Green)

oped. Sometimes modification of the cube may be on a very small scale, but crystals should always be studied carefully. For instance, a small group of colourless and transparent fluorite crystals partly coated with clear quartz crystals, figured by and in the collection of Greg and Lettsom, shows that a complex series of faces can be developed.

Other special features described from Boltsburn mine are the formation of epimorphic structures and the growth of remarkable stalactitic structures.

Epimorphs after fluorite from Boltsburn mine are often mistaken for those from the Virtuous Lady mine, Devon. These structures probably relate to several mineralising events which saw the development of the epimorphic structures late in the paragenesis. Similar fine examples of epimorphic formation have recently been described from the Brownley Hill and Rampgill mines at Nenthead.

Strange stalactitic or stalagmitic features found within the Boltsburn flats are composed principally of ring-like growths of quartz and fluorite, many with inner cores of siderite. In most of these specimens there is a small central tube. They probably formed during a later phase of metasomatism.

A feature common to Boltsburn mine and many other north Pennine localities is the partial encrustation of the cubic fluorite by quartz, pyrite or siderite. Often these are most attractive specimens. The encrusting mineral is often preferentially deposited upon certain of the cube faces, and as such provides a subject for further study and research.

Purple fluorite is the most common colour in the majority of the Weardale mines; collections of north Pennine fluorites usually have groups of fine purple interpenetrant cubes up to one inch (25 mm) across from mines such as Blackdene, Ireshopeburn, Wolf-cleugh, Groverake and the associated Frazer's Hush mine.

Within Weardale, massive light green fluorite was common, e.g. at Redburn and Groverake mines, but it rarely forms good crystal groups. Allenheads mine produced light green, purple and colourless crystals up to half an inch (12 mm) across, but perhaps the most famous coloured fluorites from this area are the vivid green interpenetrant cubic twin crystals which were obtained from Heights mine. Similar deep green crystals

were found in recent years in a small vein and associated flats in the Eastgate Cement Works Quarry. These specimens are of particular interest in exhibiting a strong colour change when exposed to strong sunlight. Within a few weeks some of the green crystals can change to pale purple. Over the last few years fine green cubic fluorite crystals have been worked from a series of small flats in Rogerley Quarry, Frosterley.

St Peter's mine, Allenheads, Northumberland is particularly famous for the groups of light green cubes with distinct yellow centres, although recent specimens collected from old workings in the mine are yellow to brownish purple crystals with siderite coatings.

Cambokeels mine produced groups of colourless or pale purple twinned cubes up to about one inch (25 mm) across.

The Scordale mines at Hilton are famous for their bright yellow, often transparent cubes. Both of the yellow fluorites from Scordale and green fluorites from Heights mines have been cut as gemstones.

Although marked colour zoning in the Boltsburn mine is rare, in other northern Pennine fluorites it is often conspicuous. Distinct single or multiple zones of purple, yellow or green, occur in some coloured crystals.

Almost all northern Pennine fluorites contain fluid inclusions, studies of which have yielded valuable information on temperatures of formation and composition of the mineralising fluids. Whereas most of these inclusions are of microscopic size, specimens of much larger inclusions, clearly visible to the naked eye, are well known in clear and often colour-zoned fluorite crystals from parts of Weardale, most notably those from Heights mine and nearby locations.

Large blocks of purple cubic crystals from the Weardale area, especially from Boltsburn mine, appear in many museum fluorescence exhibits, the crystals being caused to fluorescence by ultra-violet light. It is generally accepted that in the northern Pennines colour and fluorescence can be related to the presence of minute traces of rare earth elements. In his numerous studies of northern Pennine mineralisation the late Sir Kingsley Dunham drew attention to the distribution of colour within northern Pennine fluorite, noting that yellow is particularly common in the outer parts of the fluorite zone. He also noted that the yellow varieties typically exhibit appreciably lower rare earth contents than the purple and green forms that are so abundant and typical within the inner parts of the zone. If rare earths within the northern Pennine fluorite were derived from the Weardale Granite, it may be that they were deposited abundantly within the inner parts of the fluorite zone with much lower contents in the outer portion of the zone. The genetic significance of the rare earth content of the fluorite, in the overall context of the northern Pennine orefield, remains to be fully explored.

Fluorites from the Askrigg Block do not normally exhibit the range of colour seen further north in the Alston Block, although good greenish-yellow cubes were found at the Adelaide Level in Swaledale and some deep purple crystals were found near Cracoe. Crystals from the Greenhow area, notably some recently collected examples from Coldstones Quarry, commonly show deep purple tints on the surfaces of the crystal. Groups of colourless crystals have been collected from the Duck Street Quarry, Greenhow. These specimens commonly show a fine zonation of included sulphide, mainly tiny pyrite crystals and chalcopyrite, near the cube surface.

Outside the north Pennines, the most interesting fluorites are the light blue cubes often associated with dolomite from the orebodies at the Florence, Ullcoats and Beckermet hematite mines in west Cumbria. An indigo blue colour also occurs here, but it is rare. The colour of these blue fluorites is hard to reproduce photographically and computer enhancement techniques are often necessary.

Hematite

The iron orefields of west and south Cumbria provided huge tonnages of hematite ores that formed the basis of the heavy industries of these areas. During their lives many of the mines became famous for the quality and beauty of minerals found within these deposits. Superb specimens of barite, calcite and fluorite are discussed elsewhere, but equally famous were fine examples of varieties of hematite. Of special note are magnificent examples of the mammillated, or reniform, variety of hematite, known locally as 'kidney ore'. This occurs as bands, or as the lining of large cavities within the ore-bodies. Masses of kidney ore up to two metres across are recorded in cavities in Florence mine, Egremont. In west Cumbria many mines in the Frizington, Cleator Moor and Egremont areas were noted sources of high quality specimens, with many fine specimens coming recently from Florence mine at Egremont. Spectacular examples of kidney ore were a feature of some of the south Cumbrian mines, notably those of Roanhead and Hodbarrow. A feature of some of the kidney ore from Hodbarrow mine was the presence of curious pseudo-morphs after a cubic mineral, perhaps pyrite or fluorite. Fine examples of kidney ore were also plentiful in some of the related vein deposits of hematite within the Lake District, particularly at Nab Gill mine, Eskdale and the Knockmurton mines, near Ennerdale.

Kidney ore typically breaks into elongated conical fragments, known by the miners as 'pencil ore'. When very fine-grained this variety was well suited to the making of hematite jewellery, and substantial amounts were sold for this purpose from several mines such as Crowgarth at Cleator Moor, particularly during the 19th century.

Beautifully lustrous black crystals of 'specular hematite' were commonly encountered lining vugs in massive hematite in many deposits. Magnificent examples of this variety, commonly accompanied by large bipyramidal colourless, smoky brown or, more rarely red, quartz, are known from Florence and Haile Moor mines near Egremont.

Barium Carbonate Minerals

Witherite-Alstonite-Barytocalcite

The northern Pennine orefield is unique in the world for the abundance of the barium carbonate minerals witherite, barytocalcite and alstonite, the type localities for all of which lie within the area. Indeed, so abundant is witherite that the area was for many years the world's sole commercial source of this unusual raw material for the chemical industry.

Witherite

Witherite is relatively common and widespread in small amounts in the barium zone of the northern Pennine orefield. In places it is abundant and the area is unique in having provided the world's main commercial source of this unusual industrial mineral.

Confusion has existed in the past as to the original, or type, locality from which witherite was first described. In the first definitive account, written by Withering and read by Kirwan (published in the *Philosophical Transactions of the Royal Society*) on 22 April 1784, it was stated that 'the *Terra Ponderosa Aerata* was got out of a lead mine at Alston Moor in Cumberland'. The original specimen was obtained by Withering from his friend Matthew Boulton of Soho House, Birmingham. 'Wishing to honour Withering for his abilities and accuracy', the German mineralogist Werner named it 'witherite' in 1789.

In an account of the Anglezarke mines, Lancashire by I. Williamson, published in the *Mining Magazine*, the author discussed the type locality of witherite and notes that:

Withering was at first under the impression that his specimen came from Alston Moor, Cumberland but it was afterwards demonstrated that it was from the Anglezarke mine that the first specimen of witherite was obtained. Mining was active in the Anglezarke area of Lancashire from late 17th century to early 19th century and there is no doubt that quantities of witherite were worked. ...

The suggestion, and resulting confusion, that Withering's material derived from Anglezarke is probably due to J. Watt junior who in 1790 termed it 'aerated barites'. Watt considered that it was known on 'good authority' that no witherite had been found on Alston Moor. However, this claim is not credible and it is now accepted that the type locality is Alston Moor, although the precise locality remains unknown. It is probable that no final completely convincing solution will be possible, but as Selwyn-Turner, writing in the *Mineralogical Magazine* in 1963, concludes: 'the claims of Anglezarke rest on a more shaky foundation than that of Alston Moor.'

Dr William Withering MD FRS FLS (1741-99)

Withering was a Doctor of Medicine and a graduate of the University of Edinburgh. He joined the staff of the Birmingham General Hospital in 1779, and is best remembered for discovering the action of digitalis (foxglove) as a heart stimulant. He was an active member of the Lunar Society. This curiously named organisation was an 18th-century philosophical society whose members met for discussion, for reporting their scientific experiments, and for the exchange of associated information. It was so named as the members, never more than 14 in number, met for safety reasons, apparently related to travelling home at the time of a new moon! The meetings were held at Matthew Boulton's Soho House in Birmingham.

All members of the Lunar Society were distinguished men in their respective fields. Another noted member was the famous potter Josiah Wedgwood, who shared an interest in witherite, for Wedgwood experimented with the mineral in his various pottery glazes. Withering's famous paper on the medicinal properties of the foxglove was written in 1785 whilst he and his wife were staying at Soho House. Withering had many interests, but he was known to have some expertise in chemistry and mineralogy. There is no doubt that his most important contribution to the latter was his first definitive account of the naturally occurring form of barium carbonate – *Terra Ponderosa Aerata*.

* * *

Fresh, witherite has a greasy surface lustre, but it is often a dull white colour owing to a surface coating of barite.

Probably the finest specimens of witherite were collected at the Fallowfield mine, near Hexham where the mineral occurred in veins within Carboniferous limestones and sandstones. Crystals from here usually occur as white hexagonal bipyramids one inch (25 mm) or more across, associated with galena and sharp hexagonal bipyramids of alstonite. Fallowfield mine was worked as a commercial source of witherite, but the most famous witherite mine was Settlingtones, about five miles (8 km) to the west. Here the filling of the vein, previously worked for lead, changed almost entirely to witherite as the mine developed. Witherite production began in 1873 and continued until the mine's closure in 1969. Fine specimens of witherite were obtained from Settlingstones, although they never matched the spectacular quality of those from Fallowfield mine.

Striking hexagonal prisms of witherite were collected by Sir Arthur Russell from Nentsberry Haggs mine, Nenthead. Some fine specimens are also known from veins, which form outlying portions of the northern Pennine orefield, and which were formerly worked for barium minerals in the Durham coalfield, at New Brancepeth and South Moor.

Barytocalcite

Barytocalcite is the monoclinic polymorph of barium calcium carbonate and was originally described by Brooke in 1824, with an analysis by J. G. Children. Its type locality is the vein known as Fistas Rake at Blagill mine, Alston. The spoil heaps from this mine contain large quantities of the mineral and are today scheduled as a Site of Special Scientific Interest (SSSI). Specimens in the Russell collection, now held by the Natural History Museum, London, demonstrate that Sir Arthur Russell had obtained the mineral from a handful of other northern Pennine localities, notably Nentsberry Haggs mine, although the rather widespread occurrence of barytocalcite within the Alston Block was demonstrated by one of the present authors (B.Y.) in a paper published in 1985. Although barytocalcite may resemble witherite in its yellowish colour and greasy lustre, it is usually readily distinguished by its slender truncated monoclinic prismatic crystals.

Alstonite

When first recognised as a new mineral by James Jamieson in 1835, this was described as 'right rhombic barytocalcite' and was thus originally thought of as a variety of that mineral. The description was based on specimens from the Brownley Hill mine near Nenthead in Cumbria and Fallowfield mine near Hexham in Northumberland. It was further described in the same year by Thomas Thomson as a 'bicalcareo-carbonate of barytes'. Thomson obtained his specimen from the mineral dealer Mr John Cowper in 1834. Further publications, by both Johnston in 1837 and Thomson in 1837, presented further information with analyses. In the first of these Thomson proposed the name 'bromlite' for the mineral after the erroneous spelling of the co-type locality as Bromley Hill mine. The internationally accepted name alstonite was given by A. Breithaupt, writing in 1841, despite the apparent priority of Thomson's name 'bromlite'. Although it has long been claimed that the original specimens came from the Jug vein, exploration of the workings in the 1980s suggested that the High Cross vein was a much more likely source.

By the late 19th century specimens of this rare mineral commanded a high price. Nall, writing in 1888, noted that the workings at Brownley Hill mine which yielded alstonite had for some time been inaccessible, and that even small samples were then fetching as much as £5.00.

Alstonite has the same composition as barytocalcite, but shows orthorhombic (pseudohexagonal symmetry) whereas barytocalcite is monoclinic. Specimens from the two co-type localities can usually be told apart: alstonite crystals from the Brownley Hill mine tend to be white pseudohexagonal pyramids associated with thin hexagonal platy calcite with some pinkish barite in the matrix, whereas alstonite crystals from Fallowfield mine are usually associated with large pseudohexagonal witherite crystals. A few specimens were collected in 1909 from the barite-witherite vein at New Brancepeth Colliery, near Durham, and more recently good specimens, including some pseudomorphed by barite and calcite, from Nentsberry mine, have become well known amongst collectors.

1. ALSTONITE

From Brownley Hill mine, Nenthead, Alston, Cumbria. Aggregate of intergrown grey-white acute crystals lining a portion of a vug with calcite and witherite (max. dimension 7.5 cm). This specimen [BM 41707] was brought from Mr Alexander Bryson, in 1868. It was originally in the William Nicol FRS collection which was acquired by Alexander Bryson. This collection was sold at J. C. Stephens, 10 March 1869. The mineralogist William Nicol FRS (1768-1851) is best known for his invention of the Nicol prism; he was the first to devise a method of using thin sections of rocks and minerals for microscopic examination. (©NHMPL)

2. ALSTONITE

From Fallowfield mine, Hexham, Northumberland. Aggregate of inter-grown very pale pink acute crystals on matrix (max. dimension 10.0 cm). This specimen [BM 56336] was obtained by purchase from Mr J. Gregory, October 1855. The pink colour of some Fallowfield alstonite specimens fades on exposure to sunlight, though the cause of this is not understood. (©NHMPL)

NOTE

The fine specimens featured in this section are mostly from the collections in the Natural History Museum, London. Full locality details are given where known. However, as so often with older specimens, particularly those supplied by some 19th-century dealers, the locality information may be imprecise or incomplete with regard to the individual site or mine of origin.

4

3

3. ARAGONITE

From Frizington, Cumbria. Radiating group of cream coloured sharply pointed pyramidal crystals on a limonitic matrix (max. dimension 11.0cm). Specimens of this habit are well known from several limestone quarries and iron mines of the Frizington area. This specimen [BM 55991] was purchased from Mr F. H. Butler, in 1885. (©NHMPL)

4. ARAGONITE

From Eskett mine, Frizington, Cumbria. Radiating aggregates of grey-white tapering crystals lining a cavity in a goethite (limonite) matrix (max. dimension 25.5 cm). Obtained from Mr J. Graves of Frizington, in 1898, originally in the Couzens collection, now Russell collection [BM 1964, R 12345]. (©NHMPL)

5. ARAGONITE

From Dufton, Cumbria. White contorted stalactitic aggregate of 'flos ferri' type partially altered to calcite (max. dimension 9.5 cm). Mines of the Dufton area were known for 'flos ferri' type growth, a form rare elsewhere in the northern Pennines. This specimen [BM 91663] was part of the Allan Greg collection purchased, in 1860. (©NHMPL)

6. BARITE

From Mowbray mine, Frizington, Cumbria. Large blue crystals on 'pearl spar' dolomite (max. dimension 16.0 cm). This specimen [BM 1907, 376] was obtained by purchase from Mr J. Graves of Frizington, in 1907. (©NHMPL)

5

6

127

7

8

128

7. BARITE

From Winscales mine, Egremont, Cumbria. Group of wafer thin white tabular crystals sprinkled with reddish hematite (max. dimension 9.5 cm). This specimen [BM 1965, R 11361] was collected by Sir A. Russell, in 1930. (©NHMPL)

8. BARITE

From Pallaflat mine, Bigrigg, Egremont, Cumbria. Detached group of pale-greenish thin tabular crystals showing parallel growth and slight hematite reddening (max. dimension 11.5 cm). This specimen [BM 1913, 289] was purchased from Mr J. Graves of Frizington, in 1913. (©NHMPL)

9. BARITE

From Mowbray mine, Frizington, Cumbria. Detached pale-green doubly-terminated tabular crystal with serrated edges and thin yellow zones associated with some white dolomite (max. dimension 19.0 cm). This specimen [BM 1910, 166] was purchased from Mr J. Graves of Frizington, in 1910. (©NHMPL)

10. BARITE

From Nentsberry mine (Sun vein), Alston, Cumbria. Group of snow-white composite crystals (max. dimension 16.0 cm). This specimen [BM 1965, R 11256] was collected by Sir A. Russell, in 1930. Barite with this morphology appears to be characteristic of deposits in which barium carbonate minerals are common. (©NHMPL)

9

10

11

11. BARITE

From Settlingstones mine, Hexham, Northumberland. Group of white composite spear-shaped crystals sprinkled with a later growth of tiny barite crystals (max. dimension 11.5 cm). This crystal morphology is characteristic of barite specimens from this mine. This specimen [BM 1960, 72] was purchased from Mr W. F. Davidson of Penrith, Cumbria, in 1960. (©NHMPL)

12. BARITE

From Egremont, Cumbria. Group of pale greyish thin tabular crystals tinged with included reddish hematite; on a hematite matrix (max. dimension 13.0cm). This specimen [BM 64099] was purchased from the dealer Mr F. H. Butler in May 1889. (©NHMPL)

13. BARITE

From Mowbray mine, Frizington, Cumbria. Pair of thick tabular zoned brown ('phantom') crystals with translucent grey tips on hematite-reddened pearly dolomite (longest crystal 11.0cm). This specimen [BM 84007] was purchased from Mr J. Graves of Frizington in January 1899. (©NHMPL)

12

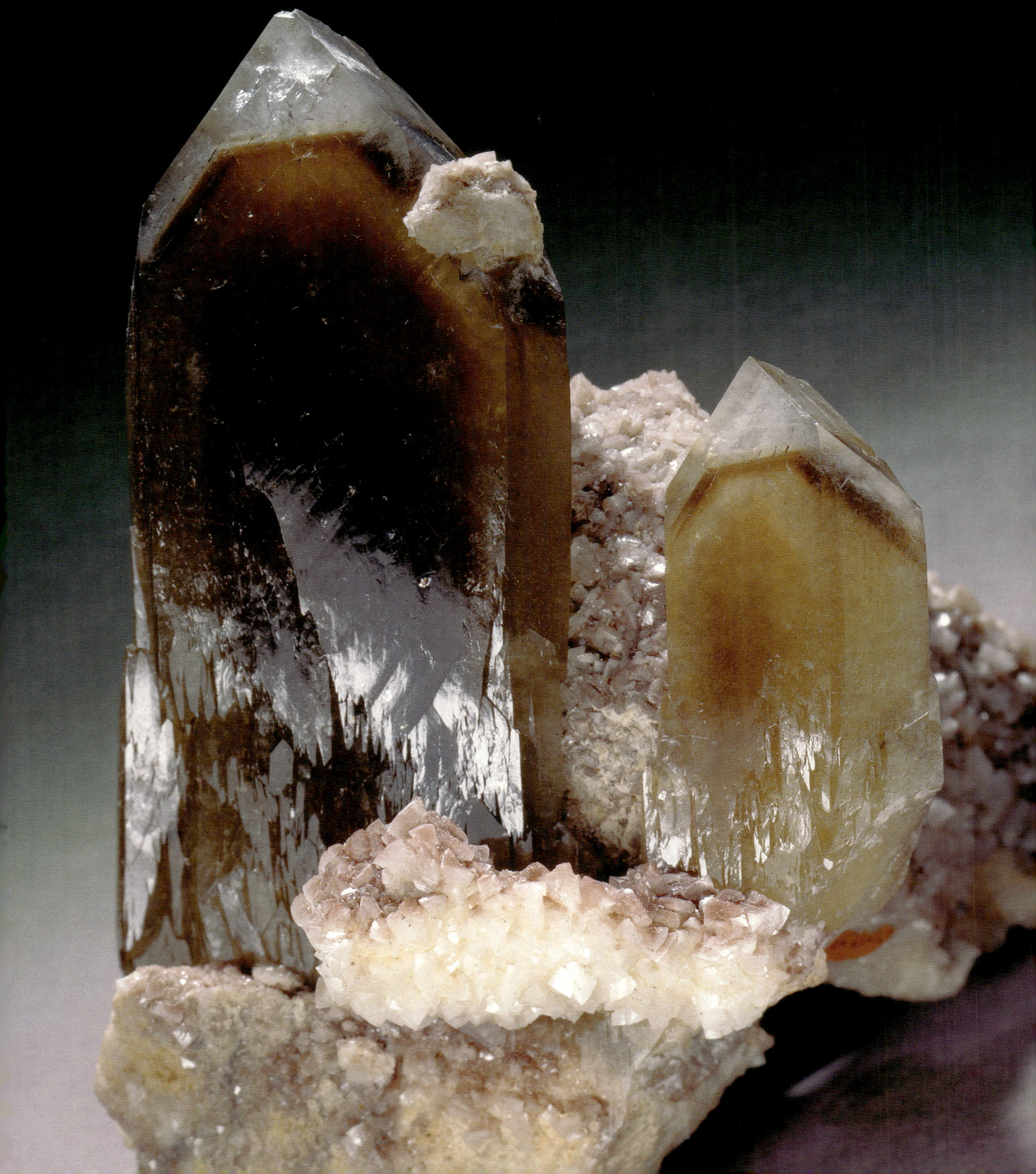

13

14 15

14. BARITE

From Parkside mine, Frizington, Cumbria. Pair of pale yellowish-green slender prismatic crystals, with smaller crystals on dolomite (longest crystal 19.5 cm). This specimen [BM 85585] was purchased from Mr J. Graves of Frizington, in 1901. As barite is such a brittle mineral, the recovery of this fine large specimen without obvious damage is a tribute to the care of its collector. (©NHMPL)

15. BARITE

From Agnes mine, Frizington, Cumbria. Detached group of pale bluish-grey translucent tabular crystals, showing incipient cleavage (max. dimension 13.0 cm). This specimen [BM 85619] was purchased from Mr J. Graves of Frizington, in 1901. (©NHMPL)

16. BARITE

From Dalmellington mine, Frizington, Cumbria. Group of amber-yellow translucent prismatic crystals on a hematite and dolomite matrix (longest crystal 19.0 cm). This specimen [BM 86819] was purchased from Mr J. Graves of Frizington, in 1904. It was figured in the Burchard and Bode publication *Mineral Museums of Europe* (1986). (©NHMPL

17. BARITE

From sump at lowest end, Wet Groves mine, Wensleydale, North Yorkshire. Group of cream coloured tabular crystals (largest crystal 5.0 cm). This specimen [cat. no. 40] was collected by Mr D. Alderson, in 1990. (©NHMPL)

132

16

18. BARITE

From Frizington, Cumbria. Group of zoned spear-shaped tabular crystals with pale amber-brown cores and white outer margins (max. dimension 11.0cm). This Russell collection specimen [BM 1964, R 11310] was originally in the Revd J. Couzens collection, 1902. (©NHMPL)

19. BARITE

From Cow Green mine, Teesdale, Co. Durham. Group of grey translucent crystals, some showing preferential overgrowths of white barite (max. dimension 20.0cm). This specimen [BM 1964, R 12344] is part of the Russell collection. Tabular crystals of this sort were characteristic of most of the barite-bearing veins of the Alston Block. (©NHMPL)

19

20. BARYTOCALCITE

From Blagill (Bleagill) mine, Alston, Cumbria. Aggregate of pale-yellow-bladed crystals lining cavities in altered limestone (max. dimension 21.0cm). This specimen [BM 41018] was purchased from the dealer Mr Burrows in 1867. This is the type locality for barytocalcite. (©NHMPL)

21. BARYTOCALCITE

From the Second Sun vein, Admiralty Concessions, Nentsberry mine, Northumberland. Stellate groups of grey-white lenticular composite crystals encrusting limestone (max. dimension 23.0cm). This specimen [BM 1964, R 6967] was collected by Sir A. Russell, in 1930. Lenticular barytocalcite crystals with this habit are characteristic of Nentsberry mine, but are uncommon elsewhere. (©NHMPL)

22. CALCITE

From Pallaflat mine, Egremont, Cumbria. Large group of colourless to grey intergrown scalenohedral crystals, some with a preferential dusting of red hematite (max. dimension 20.0cm). This specimen [BM 1907, 343] was purchased from Mr J. Graves of Frizington, in 1907. (©NHMPL)

23. CALCITE

From Bigrigg, Egremont, Cumbria. Radiating group of clear scalenohedral crystals with red hematite zoning and shallow rhombohedral terminations (max. dimension 7.0cm). This specimen [BM 1926, 1252] was part of the Dr C. O. Trechmann collection presented to the Natural History Museum, in 1926. (©NHMPL)

24

24. CALCITE

From Stank mine, Dalton in Furness, formerly in Lancashire, now in Cumbria. Group of cloudy grey modified steep rhombohedral etched crystals on botryoidal hematite (max. dimension 12.0cm). This specimen [BM 1975, 105] fluoresces pale yellowish-white under short wave U.V. light. Original collection source unknown, found in Natural History Museum, Departmental Collections. (©NHMPL)

25. CALCITE

25

From Stank mine, Dalton in Furness, formerly in Lancashire, now in Cumbria. Composite greyish translucent scalenohedron tinged pink protruding from a mass of smaller crystals and showing secondary twinned growth on alternate faces (max. dimension 16.0cm). This specimen [BM 63545] was purchased from Dr A. E. Foote of Philadelphia, USA, in April 1888. (©NHMPL)

26. CALCITE

From Egremont, Cumbria. Unusual group of heart shaped twins, filled with yellow-brown banded enclosures (limonitic), scattered on a mass of slender yellowish prismatic crystals (max. dimension 16.0cm). This specimen [BM 63661] was purchased from Mr F. H. Butler, in 1888. Specimens with this form and colouring are unusual in the west Cumbria deposits. (©NHMPL)

27. CALCITE

From Alston Moor, Cumbria. Group of grey composite scalenohedron terminated by rhombohedra in classic 'nail head' habit on galena (max. dimension 7.0cm). This specimen [BM 1972, 250] was presented by Mr P. G. Embrey, in 1972. (©NHMPL)

26

27

28. CALCITE

From Bigrigg mine, Egremont, Cumbria. Detached group of clear doubly terminated prismatic crystals with scalenohedral terminations showing a faint reddish tinge due to included hematite (max. dimension 8.0cm). This specimen [BM 85583] was obtained by purchase from Mr J. Graves of Frizington, in 1901. (©NHMPL)

29. CALCITE

From Bigrigg mine, Egremont, Cumbria. Sparkling aggregate of clear prismatic crystals with scalenohedral and rhombohedral terminations and minor inclusions of reddish hematite (max. dimension 19.0cm). This specimen [BM 85698] was purchased from Mr J. Graves of Frizington, in 1902. (©NHMPL)

28

29

30. CALCITE

From Bigrigg mine, Egremont, Cumbria.
Confused aggregate of clear sharp rhombo-
hedrally terminated prismatic crystals with
inclusions of black manganese oxides
(longest crystal 15.4 cm). This specimen
[BM 86419] was purchased from Mr J.
Graves of Frizington, in 1903. Used as
figure 6, plate 111 in A. L. Tutton's book
Crystals, published in 1911, and figured
Calcite in P. Bancroft's book *The World's
Finest Minerals and Crystals* (1973).
(©NHMPL)

31. CALCITE

From Bigrigg mine, Egremont, Cumbria.
Aggregate of colourless prismatic crystals
with scalenohedral and rhombohedral terminations, on a limonitic matrix (max.
dimension 14.0 cm). This specimen [BM
85707] was obtained by purchase from Mr
J. Graves of Frizington, in 1902. (©NHMPL)

30

31

32

33

142

32. CALCITE

From Alston Moor, Cumbria. Group of pale greyish-yellow translucent rough prismatic crystals with slightly domed terminations on a matrix of quartz and fluorite (max. dimension 14.0 cm). This specimen [BM 90633] was part of the Allan-Greg collection, purchased in 1860. (©NHMPL)

33. CALCITE

From Stank mine, Dalton-in-Furness, formerly in Lancashire, now in Cumbria. Detached stellate (radiating) group of reddish-brown scalenohedral crystals (max. dimension 9.5 cm). This specimen [BM 1985, MI 5341] was purchased from the dealer F. H. Butler, in October 1885. (©NHMPL)

34. CALCITE

From Rodderup (sometimes spelt Rotherhope) Fell mine, Alston, Cumbria. Unusual short-prismatic greyish-white and darker-grey bi-coloured twinned crystals on a mass of smaller prismatic crystals (max. dimension 11.0 cm). This specimen [BM 1964, R 5568] is part of the Sir A. Russell collection. A series of such specimens was collected in the 1930s, many growing directly on galena. (©NHMPL)

35. CALCITE

From Egremont, Cumbria. Doubly terminated pink to colourless scalenohedral crystal [BM 85627] (max. dimension 8.5 cm). Purchased from Mr J. Graves of Frizington, Cumbria, in November 1901. (©NHMPL)

34

35

143

36

37

36. CALCITE

From Egremont, Cumbria. Detached, pale pinkish to clear prismatic crystals forming a right angled twin (max. dimension 10.0 cm). This specimen [BM 63673] was purchased from Mr F. H. Butler, in 1888. Francis H. Butler (1849-1935) was a mineral and fossil dealer, and executor to Richard Talling's collection of minerals. (©NHMPL)

37. CALCITE

From Egremont, Cumbria. Detached clear 'heart shaped' twin dusted with reddish hematite and sharing a prismatic continuation of one part of the crystal (max. dimension 10.0 cm). This specimen [BM 85703] was purchased from Mr J. Graves of Frizington, in 1902. (©NHMPL)

38. CALCITE

From Cumbria (probably Croft Pit, Bigrigg). Grey, translucent complex twinned crystal containing black manganese oxide inclusions (max. dimension 6.5 cm). This specimen in the Russell collection [BM 1964, R 5499] was originally in the Revd Couzens collection and purchased from Mr J. Graves of Frizington, in 1902. Original Couzens label records 'new variety – vary rare!' (©NHMPL)

39. CALCITE

From Bigrigg mine, Egremont, Cumbria. Detached pale-brown translucent 'butterfly twin' containing phantoms of earlier crystal growth (max. dimension 7.0 cm). This specimen [BM 1985, MI 8328] was obtained by purchase from Mr J. Graves of Frizington, in 1902. (©NHMPL)

40. CALCITE

From Croft Pit, Bigrigg, Cumbria. Detached clear 'heart shaped' crystal (max. dimension 9.5 cm). This specimen [BM 1964, R 5613] was obtained from the sale of Colonel Rimington's collection, by auction in Paris, 1912, Lot 542, Colonel J. W. Rimington (1832-1909). This collection of minerals was sold by J. C. Stevens in 1891, and also in Paris in 1912. (©NHMPL)

41. CALCITE

From Stank mine, Dalton-in-Furness, Cumbria. Detached clear complex 'heart shaped' twinned crystal, containing faint phantoms of earlier crystal growth (max. dimension 5.5 cm). This specimen [BM 1964, R 5773] was originally in the collection of Lady Constance Russell (mother of Sir Arthur Russell). (©NHMPL)

42. CALCITE

From Bigrigg mine, Egremont, Cumbria. Detached pale-brown translucent 'butterfly twin' containing distinct red hematite coated phantoms (max. dimension 5.0 cm). This specimen [BM 1964, R 5604] is in the Russell collection. (©NHMPL)

43. CALCITE

Simply labelled 'Cumbria', but almost certainly from the Egremont area. Detached, water clear 'heart shaped' twin crystal showing an unusually well defined twin plane and rippled prism faces (max. dimension 9.5 cm). This specimen [BM 1964, R 5500] was originally in Revd Couzens' collection and purchased from Mr J. Graves of Frizington, in 1909. Couzens label reads 'very fine, very rare!' (©NHMPL)

41

42

43

147

44

45

148

44. CERUSSITE

From Stanhopeburn mine, Weardale, Co. Durham. Sprays of radiating slender white prisms on a limonitic matrix. Upper levels of this mine were the source of numerous fine specimens of this sort (max. dimension 13.0 cm). Specimen [BM 1920, 371] obtained by purchase from J. R. Gregory & Co., in 1920. Although a common mineral in the northern Pennines, Stanhopeburn was one of a very few mines that yielded fine specimens of this sort. (©NHMPL)

45. CHALCOPYRITE

From Greenside mine (station 21, 150 fathoms level), Glenridding, Patterdale, Cumbria (formerly Westmorland). Dull brassy metallic crystals on grey drusy quartz (max. dimension 10.0 cm). This specimen [BM 1964, R 5366] was collected by Sir A. Russell, 1938. Well-crystallised chalcopyrite is uncommon in the Lake District. (©NHMPL)

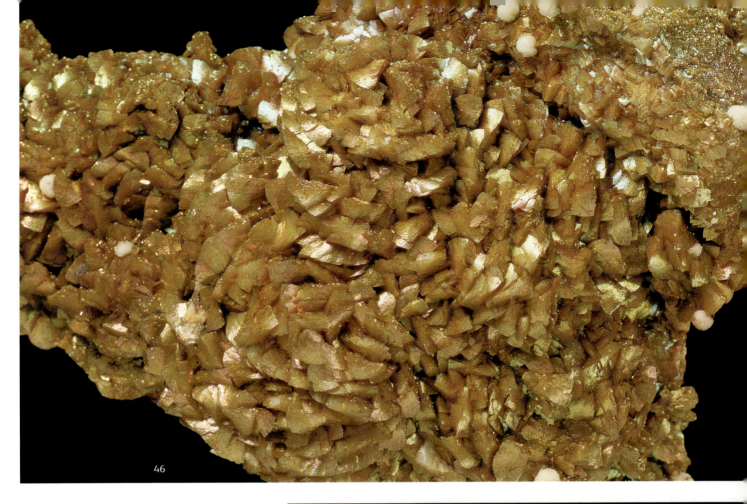

46

46. DOLOMITE

From Frizington, Cumbria. Crust of typical pale brown 'curved' saddle-rhombs, exhibiting slight iridescence (max. dimension 14.0cm). Such specimens were common in this area. The iridescence is due to coatings of microscopic tarnished pyrite crystals [BM 82709]. By purchase from Mr J. Graves of Frizington, in 1897. (©NHMPL)

47. DOLOMITE

From Slitt vein, Blackdene mine, Ireshopeburn, Weardale, Co. Durham. Aggregates of white pearly curved crystals coating grey limestone with some brown siderite [BM 1980, 40] (max. dimension 13.0cm). By purchase from the dealer Mr R. W. Barstow, in 1980. (©NHMPL)

47

149

48

49

48. FLUORITE

From Weardale, Co. Durham. No other locality details are recorded. Group of pale-yellow sharp cubic crystals with distinct lilac edges on a matrix of buff dolomite (max. dimension 8.0cm). This specimen [BM 1907, 522] was obtained by bequest of Miss Caroline Birley, in 1907. (©NHMPL)

49. FLUORITE

From Boltsburn mine, Rookhope, Weardale, Co. Durham. Group of intergrown sharp lilac cubic crystals with slightly darker lilac edge zones with a little buff dolomite or ankerite (max. dimension 13.0cm). This specimen [BM 1985, MI 7398] is a typical habit of fluorite from this mine. (©NHMPL)

50. FLUORITE

From Boltsburn mine, Rookhope, Weardale, Co. Durham. Unusual epimorphs of brown siderite after cubic fluorite crystals with some remnant purple fluorite (max. dimension 22.0cm). This specimen [BM 1964, R 1234] was collected by Sir A. Russell, in 1926. Such specimens, which were rare in the northern Pennines, resemble the hollow cubic epimorphs, known as 'boxes', formerly found at Virtuous Lady mine in Devon. (©NHMPL)

51. FLUORITE

From East Arn Gill mine (Adelaide Level), Muker, Wensleydale, North Yorkshire. Group of pale yellow sharp cubic crystals, with distinct edge modifications and included zones of tiny chalcopyrite crystals with cerussite coated galena crystals (max. dimension 11.0cm). This specimen [BM 1964, R 1406], part of the Russell collection, was collected by Russell in 1920. Coloured fluorite crystals are generally uncommon in the Askrigg Pennines where most of the fluorite is white or colourless. (©NHMPL)

50

51

151

52. FLUORITE

White's Level, Westgate, Weardale, Co. Durham. Deep green twinned cubic crystals, with narrow edge modifications and showing natural purplish fluorescence on an iron stained limestone matrix (max. dimension 23.0cm). This specimen, part of the Russell collection [BM 1964, R 1524] was bought from the dealer S. Henson. (©NHMPL)

53. FLUORITE

From Boltsburn mine, Rookhope, Weardale, Co. Durham. Group of intergrown sharp, opaque cubes on a sandy dolomite matrix (max. dimension 18.0cm). This specimen [BM 1964, R 1485], part of the Russell collection, was obtained from T. Reed, in 1930. (Thomas Reed, a contemporary of Sir Arthur Russell, was foreman of Boltsburn mine). (©NHMPL)

53

54. FLUORITE

From Heights mine, Westgate, Weardale,
Co. Durham. Group of deep-green twinned
sharp cubic crystals with narrow edge modifications and showing natural purple-blue
fluorescence (max. dimension 11.5 cm). This
specimen [BM 1964, R 1526], part of the
Russell collection, was purchased from S.
Henson, the mineral dealer, 1848-1930.
(©NHMPL)

52

54

153

55. FLUORITE

From St Peter's mine, East Allendale, Northumberland. Aggregate of intergrown clear green cubic crystals, some with yellow centres associated with silvery metallic galena (max. dimension 26.0 cm). This specimen [BM 1964, R 4729] is part of the Russell collection. This unusual colour appears to be unique to this locality. (©NHMPL)

56. FLUORITE

From Scordale mine, Dow Scar, Hilton, Appleby, Cumbria (formerly Westmorland), from High Level flat in the Jew Limestone (max. dimension 5.0 cm). Detached bright-yellow translucent cube showing facial growth striae and very thin edge modifications. This specimen [BM 1964, R 1731] is part of the Russell collection. (©NHMPL)

57. FLUORITE

From Rodderup (sometimes spelt Rotherhope) Fell mine, Alston, Cumbria. Purple, dark blue-violet and honey yellow banded and in part yellowish-white mottled interpenetrating twinned cubes; small crystals of calcite and quartz on black silicified limestone (max. dimension 16.0 cm; individual cubes up to 3.5 cm). This specimen [BM 1965, R 1695] was obtained from H. Millican, in 1930. (©NHMPL)

58

58. FLUORITE

From Weardale, Co. Durham (probably from Boltsburn mine, Rookhope, Weardale). Deep mauve twinned cubic crystal with distinct edge modifications and vicinal faces, on a quartz matrix and associated with white lenticular calcite crystals (max. dimension 15.0 cm). This specimen [BM 31360] was purchased from Mr B. Wright, in 1860. (©NHMPL)

59. FLUORITE

Zinc flat, Slitt vein, Cambokeels mine, Eastgate, Weardale, Co. Durham. Aggregate of intergrown and twinned colourless to white sharp cubic crystals showing clear vicinal faces (max. dimension 9.0 cm). This specimen [RFS 1] was originally in the L. and P. Greenbank collection, in 1990. Colourless fluorite is rather uncommon in the Alston Pennines, though several good specimens were collected from the deeper levels of this mine. (R. F. Symes)

59

60. FLUORITE

From Weardale, Co. Durham. Group of deep purple intergrown sharp cubic crystals with white cloudy centres (max. dimension 20.0 cm). This specimen [BM 1968, 480] was purchased from Earth Science Imports (F. L. Smith and C. L. Key collection) of New Jersey, in 1968. (©NHMPL)

61

62

61. FLUORITE

From Boltsburn mine, Rookhope, Weardale, Co. Durham. Group of intergrown pale greenish cubes with yellowish cores, with rosettes of brown siderite sprinkled with sphalerite (max. dimension 16.0cm). This specimen [BM 1964, R 1495] is part of the Russell collection, and was obtained from T. [Thomas] Reed, in 1930. A later partial coating of siderite crystals is a common feature of many Boltsburn fluorite specimens. (©NHMPL)

62. FLUORITE

From Duck Street Quarry, Greenhow, Pateley Bridge, North Yorkshire. Group of mostly colourless translucent cubic crystals with parallel inclusions of tiny chalcopyrite crystals appearing as faint grey-green edge zones (max. dimension 13.0cm). This specimen [BM 1964, R 1856], part of the Russell collection, was collected by A. and M. Russell in 1932. (©NHMPL)

63. FLUORITE

From Burtree Pasture mine, Cowshill, Weardale, Co. Durham. Group of greenish cubic crystals with naturally fluorescent purple edge zones, with some white shallow rhombohedral calcite crystals (max. dimension 11.0cm). This specimen [BM 1964, R 1557], part of the Russell collection, was originally in the J. P. Walton collection (John P. Walton, 1838-1915) of Acomb High House, Hexham, Northumberland. (©NHMPL)

64. GALENA

From Nenthead, Cumbria. Dull grey metallic composite cubic crystals on matrix with black sphalerite and orange-brown ankerite crystals (max. dimension 36.0cm). This specimen [BM 1985, MI 7199] was obtained by purchase from Mr J. Graves of Frizington. (©NHMPL)

65. GALENA

From Boltsburn mine, Rookhope, Weardale, Co. Durham. Group of bright silvery-grey metallic cubo-octahedral crystals with honey-brown ankerite or siderite crystals (max. dimension 9.5cm). Collected by Thomas Reed, mine foreman, in 1927. This specimen [BM 1964, R 3811] is in the Russell collection. (©NHMPL)

66. GALENA

From Allenheads mine, Northumberland. Group of dull-grey rough cubic crystals showing successive depressed etching of faces, with bluish drusy quartz (max. dimension 10.0cm). This structure was known to the miners as 'seals'. This unusual specimen [BM 1985, MI 15332] was presented by Sir W. W. Smyth FRS (1817-1890). (©NHMPL)

66

65

67

68

162

67. HAUSMANNITE

From Wyndham Pit, Bigrigg, Egremont, Cumbria. Group of sharp twinned black crystals with hematite (max. dimension 8.0cm). This specimen [BM 1919, 229] was presented by F. N. Ashcroft, in 1919. Many fine examples of this mineral were collected from the small manganese-rich pocket of ore found here in association with hematite. (©NHMPL)

68. GRAPHITE

From Borrowdale, Cumbria. Black shiny foliated crystal aggregates (max. dimension 11.0cm). This specimen [BM 1964, R3] is part of the Russell collection. (©NHMPL)

69. HEMATITE ('kidney ore')

From Cleator Moor, Cumbria. Deep red-brown mammillated aggregate (max. dimension 13.5 cm). This specimen [BM 28265] was obtained by purchase from Mr Wright, in 1859. (©NHMPL)

70. HEMATITE

From Cumbria. Reddish-brown botryoidal mass of 'kidney ore' coated with drusy specular hematite, with grey bipyramidal quartz crystals (max. dimension 28.0 cm). This specimen [BM 1985, MI 5977] was purchased from Mr F. S. Butler, in 1890. (©NHMPL)

69

70

163

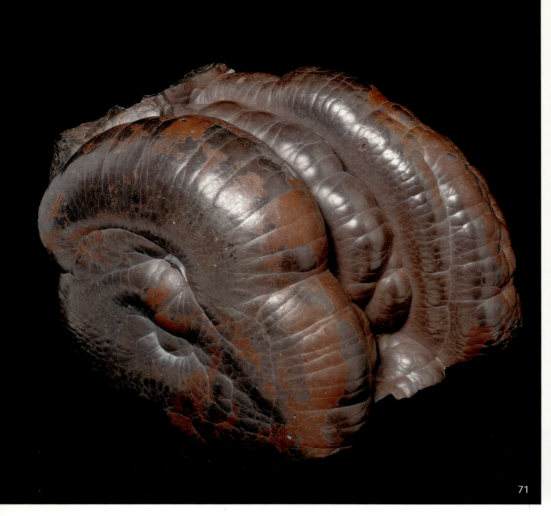

71

72

71. HEMATITE ('kidney ore')

From Nigel mine, Furness, formerly in Lancashire, now in Cumbria. Reddish and reddish-brown mammillated mass with furrowed surface (max. dimension 24.0 cm). This specimen [BM 1985, MI 27277] was presented by Sir K. C. Dunham, in 1939. 'Kidney ore' with this very distinctive appearance was once a feature of the Nigel mine. (©NHMPL)

72. MILLERITE

From Boltsburn mine, Rookhope, Weardale, Co. Durham. Detached divergent group of brassy metallic slender crystals (max. dimension 3.5 cm). This specimen [BM 1964, R 4083] was collected by Sir A. Russell, in 1926. Although found in tiny amounts at a handful of northern Pennine locations, specimens of this size and quality are exceptional. (©NHMPL)

73. NICKELINE

From the 70 fathom level, Settlingstones mine, Fourstones, Northumberland. Pale pinkish metallic aggregates in matrix (sawn and polished specimen) (max. dimension 17.5 cm). This specimen [BM 1964, R 4090] was collected by Sir A. Russell, in 1925. Although about half a tonne of this nickel ore was raised at Settlingstones in the 1920s, nickeline is a rare mineral in the northern Pennines. (©NHMPL)

74. PECTOLITE

From Copt Hill Quarry, Cowshill, Weardale, Co. Durham. Pale greenish-white radiating spherulitic aggregates in matrix with some grey calcite (max. dimension 23.0 cm). This specimen [BM 86416] was purchased from Mr J. Graves of Frizington, in 1903. Fine examples of pectolite are found coating joints in the Whin Sill of the northern Pennines. (©NHMPL)

73

74

165

75. PYROMORPHITE

From Grass Hill mine, Teesdale, Co. Durham. Pale green compact, intergrown crystalline mass with black manganese oxides (max. dimension 9.0 cm). This specimen [BM 1964, R 7998] was collected by Sir A. Russell, in 1930. Such rich examples of pyromorphite are rare in the northern Pennines. (©NHMPL)

76. PYROMORPHITE

From Old Brandlehow mine, near Keswick, Cumbria. Crust of green prismatic crystals on quartz (max. dimension 11.0 cm). This specimen [BM 1958, 127] was collected and presented by A. W. G. Kingsbury, in 1953, from the top level on the east side of Catbells, Derwentwater. Although many of Kingsbury's 'finds' are now discredited, examples of similar pyromorphite have been obtained by many collectors from this locality. (©NHMPL)

77. QUARTZ

From Frizington, Cumbria. Dark smoky brown translucent bipyramidal crystal intergrown with black specular hematite (max. dimension 10.5 cm). This specimen [BM 1911, 212] was obtained by purchase from Mr S. Henson, in 1911. Such smoky crystals were once comparatively common in many of the west Cumbrian iron ore mines. (©NHMPL)

76

77

78

79

78. QUARTZ WITH HEMATITE ('eisenkiesel')

From Cleator Moor, Cumbria. Aggregates of clear glassy crystals with orange-brown hematite inclusions on black specular hematite (max. dimension 20.0cm). This specimen [BM 41165] was purchased from Mr B. M. Wright, in 1867. (©NHMPL)

80

79. QUARTZ

From Alston, Cumbria. Intergrown greyish glassy crystals partly encrusting greyish-lilac cubic fluorite crystals with pitted surfaces (max. dimension 15.0cm). This specimen [BM 58407] is believed to be from the Sir Hans Sloane collection, bequeathed to the nation in 1753. Many fine specimens of northern Pennine fluorite exhibit complete or partial coatings of similar white quartz crystals. (©NHMPL)

80. QUARTZ

From Frizington, Cumbria. Group of smoky brown bipyramidal crystals with red hematite inclusions on black specular and red-brown massive hematite (max. dimension, 15.0cm). This specimen [BM 1985, MI 7919] was purchased from Mr J. Graves of Frizington, in 1902. (©NHMPL)

81

81. QUARTZ

A fine cut and polished 'fortification agate' from the lavas of the Cheviot Hills. This specimen (max. dimension 4.5 cm) is in the private collection of Mr Alan Pringle, who collected it from glacial deposits near Quickeningcote in the Coquet valley, Northumberland. (©NERC)

82. RHODOCHROSITE

From Wyndham Pit, Bigrigg, Egremont, Cumbria. Bright pink rhombohedral crystals and massive, lining a cavity in manganese oxide bearing matrix (max. dimension 7.0cm). Part of the Russell collection [BM 1964, R 6209], collected by Russell in 1930. (©NHMPL)

82

83. SMITHSONITE

From Farnberry mine, Alston, Cumbria. Bright-yellow stalactitic and botryoidal aggregates on yellow massive smithsonite and on a goethite matrix (max. dimension 17.0cm). Originally in J. P. Walton collection, 1848, now in Russell collection. Although smithsonite is common and widespread in the northern Pennines, it is usually found as the dull brown cellular variety known as 'dry bone'. Beautiful crystalline masses of this sort are generally rare, but seem once to have been fairly common at this location. [BM 1964, R 6229]. (©NHMPL)

84. SPHALERITE

From Nenthead, Alston, Cumbria. Aggregate of black lustrous twinned composite crystals on limestone matrix (max. dimension 9.0cm). This specimen [BM 1909, 450] was obtained by purchase from Mr J. Graves of Frizington, in 1909. Black lustrous crystals of this sort are characteristic of much of the sphalerite in the northern Pennines. (©NHMPL)

85. SPHALERITE

From Boltsburn mine, Rookhope, Weardale, Co. Durham. Aggregate of brownish-black lustrous twinned composite crystals with some purple fluorite (max. dimension 15.0cm). This specimen [BM 1909, 518] was obtained from Mr W. W. Bell of Rookhope, Weardale, 1909. (©NHMPL)

86. SPHALERITE

From Alston Moor, Cumbria. Black lustrous composite crystals on creamy ankerite with pale purple fluorite (max. dimension 14.0cm). This specimen [BM 41581] was purchased from Mr Burrows, in 1868. (©NHMPL)

85

87. STRONTIANITE

From Pateley Bridge, North Yorkshire. Groups of sharp grey-white pyramidal crystals on lenticular 'cock's-comb' barite crystals on matrix [BM 35334] (max. dimension 9.0cm). Presented by Dr Murray of Scarborough, 1863. Strontianite is known from several mines in the Askrigg Pennines, though specimens of this quality are rare. (©NHMPL)

88. WITHERITE

From Fallowfield mine, Hexham, Northumberland. Greyish white bipyramidal pseudo-hexagonal crystals with alstonite and calcite (max. dimension 6.0cm). This specimen [BM 62914] was purchased from the dealer Mr P. Gilmore, in 1887. Figured in H. A. Miers, *Mineralogy*, 1929. Many fine examples of Fallowfield witherite are accompanied by alstonite crystals. (©NHMPL)

89. WITHERITE

From Cox's vein, Nentsberry mine, Alston, Cumbria. Grey-white bipyramidal pseudo-hexagonal crystal on silvery-grey galena and creamy dolomite (max. dimension 8.0cm). This specimen [BM 1964, R 6627] is part of the Russell collection, obtained from Percy Blight, 1933. (©NHMPL)

90. WITHERITE

From Arkengarthdale, Reeth, North Yorkshire. Groups of creamy opaque bipyramidal pseudohexagonal crystals on matrix [BM 89681] (max. dimension 8.5cm). Originally part of the Allan-Greg collection, purchased in 1860. Although witherite is a common mineral in the Askrigg Block, specimens of this quality are uncommon. (©NHMPL)

89

90

175

91

91. WITHERITE

From Nentsberry mine, Alston, Cumbria. Group of snow-white pseudohexagonal prismatic crystals with porous surfaces and partly altered to barite (largest crystal 3.3 cm). This specimen [BM 1964, R 6641] was collected in 1931 by Sir A. Russell. It was collected from a large cavity off the Liverick vein in the cross-cut between the Treloar and High Raise vein. The specimen figured in the Peter Bancroft publication *The World's Finest Minerals and Crystals*, 1973. (©NHMPL)

92. WITHERITE

From New Brancepeth Colliery, Co. Durham. Aggregates of pale-yellow composite sharply terminated pseudohexagonal crystals on matrix (max. dimension 22.0 cm). This specimen [BM 1964, R 6703] was collected by Sir A. Russell, in 1943. (©NHMPL)

92

Selected Bibliography

IT is possible here to cite only a selection of the technical and historical references which touch upon the extremely varied geology, mineralogy and long history of mining in the area covered by this volume. Although long, the list does not pretend to be comprehensive and, in a work of this sort, some omissions are always inevitable.

All references are to published works that may be accessed comparatively readily. Although a number of unpublished MSc and PhD theses, technical reports and society newsletters may contain interesting information on certain sites and topics, these are generally difficult to access and have therefore not been included.

References to the area's geology focus upon some of the most significant 'milestone' research papers, together with important local or regional syntheses, most notably the *Memoirs* and other publications of the British Geological Survey. Within all of these may be found substantial lists of additional references on specific areas or topics. Within this list, priority has been given to those references that deal specifically with mineral occurrences, or with mineralisation and metallogenic processes. The published literature on collectors and collecting is comparatively meagre for much of the area, though the most significant references are included. A selection of the most important texts relating to mining history and, where appropriate, mineral technology, are also included.

Geological maps at a variety of scales, and related publications of the British Geological Survey, may be obtained through booksellers or directly from the British Geological Survey, at the Kingsley Dunham Centre, Keyworth, Nottingham NG12 5GG (website www.bgs.ac.uk).

Topographical maps are produced and published by the Ordnance Survey and may be obtained through booksellers.

ADAMS, J. (1988): *Mines of the Lake District Fells* (Clapham [via Lancaster]: Dalesman Books).

ADDISON, P. L. (1889-90): 'Description of the Cleator Iron Company's barytes and umber mines and refining mills in the Caldbeck Fells', *Minutes of Proceedings of the Institution of Civil Engineers*, vol. 102, pp. 283-91.

AKHURST, M. C., R. R. CHADWICK, R. W. HOLLIDAY, M. McCORMAC, A. McMILLAN, D. MILLWARD and B. YOUNG (1997): 'Geology of the west Cumbria district', *Memoir of the British Geological Survey, England and Wales*, Sheets 28, 37 and 47.

ALMOND, J. K. (1977): 'The Nenthead and Tynedale Lead and Zinc Company Ltd', *Memoirs of the Northern Mines Research Society, British Mining*, no. 5, pp. 26-40.

ALMOND, J. K. (2003): 'Dressing north Pennine fluorite ores for the markets', pp. 73-129 in R. A. FAIRBAIRN (ed.): *Fluorspar in the Pennines* (Killhope, Co. Durham: Friends of Killhope).

ALMOND, J. K. and H. L. BEADLE (1979): 'A guide to the past lead industry in Swaledale and Weardale', *Cleveland Industrial Archaeological Society Research Report*, no. 2.

ANDERSON, W. (1945): 'On the chloride waters of Great Britain', *Geological Magazine*, vol. 82, pp. 267-74.

ANON (1940): *Witherite and its industrial uses,* The Holmside and South Moor Collieries Ltd and the owners of Settlingstones Mines Ltd.

APPLETON, J. D. and A. J. WADGE (1976): 'Investigations of tungsten and other mineralization associated with the Skiddaw Granite near Carrock Mine, Cumbria', *Institute of Geological Sciences Mineral Reconnaissance Programme*, Report no. 7.

ARTHURTON, R. S. (1971): 'The Permian evaporites of the Langwathby Borehole, Cumberland', *Report of the Institute of Geological Sciences*, no. 71/7.

ARTHURTON, R. S., I. C. BURGESS and D. W. HOLLIDAY (1978): 'Permian and Triassic', pp. 189-206 in F. MOSELEY (ed.): *The Geology of the Lake District*, Yorkshire Geological Society, Occasional Publication, no. 3.

ARTHURTON, R. S. and J. E. HEMINGWAY (1972): 'The St Bees Evaporites – a carbonate-evaporite formation of Upper Permian age in West Cumberland, England', *Proceedings of the Yorkshire Geological Society*, vol. 38, pp. 565-92.

ARTHURTON, R. S., E. W. JOHNSON and D. J. C. MUNDY (1988): 'Geology of the country around Settle', *Memoir of the British Geological Survey*.

ARTHURTON, R. S. and A. J. WADGE (1981): 'Geology of the country around Penrith', *Memoir of the Geological Survey of Great Britain*.

ASHBURN, J. (1963): 'Mining witherite in N. W. Durham', *Colliery Guardian*, vol. 207, pp. 269-76.

AVELINE, W. T. (1873): 'The geology of the southern part of the Furness district in north Lancashire', *Memoir of the Geological Survey of Great Britain*.

AVELINE, W. T. and T. McK. HUGHES (1888): 'Geology of the country around Kendal, Sedbergh, Bowness and Tebay', *Memoir of the Geological Survey of Great Britain, England and Wales*, Sheet 39 (2nd edition), revised by A. STRACHAN.

BACK TRACK (2003): 'W. H. Auden on Railways', *Back Track (Railway Magazine)*, vol. 17, no. 8.

BALL, T. K., N. J. FORTEY and T. J. SHEPHERD (1985): 'Mineralisation at the Carrock Fell tungsten mine, N. England: paragenetic, fluid inclusion and geochemical study', *Mineralium Deposita*, vol. 20, pp. 57-65.

BANCROFT, P. (1973): *The World's Finest Minerals and Crystals* (Thames and Hudson).

BANCROFT, P. (1984): 'Gem and Crystal Treasures. Western Enterprises', *Mineralogical Record*.

BANNISTER, F. A., M. H. HEY, and G. F. CLARINGBULL (1950): 'Connellite, buttgenbachite and tallingite', *Mineralogical Magazine*, vol. 29, pp. 280-86.

BARBER, R. (1976): *Iron Ore and After. Cleator Moor 1840-1960* (York University in association with Cleator Moor Local Studies Group).

BARCLAY, W. J., N. J. RILEY and G. E. STRONG (1994): 'The Dinantian rocks of the Sellafield area, West Cumbria', *Proceedings of the Yorkshire Geological Society*, vol. 50, pp. 37-49.

BARKER, J. L. (1972): 'The lead miners of Swaledale and Arkengarthdale in 1851', *Memoirs of the Northern Cavern and Mines Research Society*, vol. 2, pp. 89-97.

BARNES, J. (1992): 'The Mineral Collection of John Ruskin in the Ruskin Gallery, Sheffield', *U.K. Journal of Mines and Minerals*, no. 11, pp. 26-29.

BARNES, R. P., K. AMBROSE, D. W. HOLLIDAY and N. S. JONES (1994): 'Lithostratigraphical subdivision of the Triassic Sherwood Sandstone Group in west Cumbria', *Proceedings of the Yorkshire Geological Society*, vol. 50, 51-60.

BASHAM, I. R., T. K. BALL, B. BEDDOE-STEPHENS and U. McL. MICHIE (1982): 'Uranium-bearing accessory minerals and granite fertility: II Studies of granites from the British Isles', pp. 398-413, in *Symposium on uranium exploration methods* (International Atomic Energy Agency [IAEA]).

BATESON, J. H. and A. D. EVANS (1982): 'Mineral exploration in the Ravenstonedale area', *Institute of Geological Sciences Mineral Reconnaissance Report*, no. 57.

BATESON, J. H. and C. C. JOHNSON (1983): 'A mineral reconnaissance of the Dent-Ingleton Area of the Askrigg Block, Northern England', *Institute of Geological Sciences Mineral Reconnaissance Report*, no. 64.

BATESON, J. H., C. C. JOHNSON and A. D. EVANS (1983): 'Mineral reconnaissance in the Northumberland Trough', *Institute of Geological Sciences Mineral Reconnaissance Report*, no. 62.

BEADLE, H. L. (1968): 'The history of Cowgreen Mines', *Bulletin of the Durham Local History Society*, no. 9, pp. 1-10.

BEADLE, H. L. (1977): 'Lady's Rake Mine, Harwood-in-Teesdale', *Cleveland Industrial Archaeologist*, no. 7, pp. 17-23.

BEADLE, H. L. (1980): 'Mining and smelting in Teesdale', *Cleveland Industrial Archaeological Society Research*, Report no. 3.

BEAN, S. (1737): *Rara Avis in Terris: or, The Laws and customs of the lead mines within the mineral liberty of Grassington cum Membris ... Leeds.*

BECKERMET MINE (1990): Reprint by the Friends of Whitehaven Museum in association with the West Cumbria Mines Research Group.

BEDDOE-STEPHENS, B. and N. J. FORTEY (1981): 'Columbite from the Carrock Fell tungsten deposit', *Mineralogical Magazine*, vol. 44, pp. 217-23.

BEDDOE-STEPHENS, B. and I. MASON (1991): 'The volcano-genetic significance of garnet-bearing minor intrusions within the Borrowdale Volcanic Group, Eskdale area, Cumbria', *Geological Magazine*, vol. 128, pp. 505-16.

BEVINS, R. E., (1994): *A mineralogy of Wales* (Cardiff: National Museum of Wales), Geological series, no. 16.

BEVINS, R. E., J. H. OLIVER and L. J. THOMAS (1985): 'Low grade metamorphism in the paratectonic Caledonides of the British Isles', *Earth Evolution Sciences Monograph Series*, no. 1, pp. 57-79.

BGS = BRITISH GEOLOGICAL SURVEY (see below).

BINNEY, E. W. (1868): 'On the age of the haematite iron deposits of Furness', *Memoir of the Literary and Philosophical Society of Manchester*, vol. 7, pp. 55-61.

BIRD, R. H. (1980): 'The Mines of Grassington Moor and Wharfedale', *British Mining* (Northern Mines Research Society).

BIRLEY, R. and B. YOUNG (1993): 'Vivianite of post-Roman origin from Vindolanda, Northumberland', *Transactions of the Natural History Society of Northumbria*, vols 56, 65-67.

BLANCHARD, I. (1992): 'Technical implications from silver to lead smelting in 12th-century England', pp. 15-17 in *Boles and Smelt Mills Seminar*, Historical Metallurgical Society, 15-17 May 1992.

BLANCHARD, I. (2001): *Mining, metallurgy and minting in the middle ages* (Stuttgart: Steiner).

BLAND, D. J. (1988): 'An occurrence of silver and gold in Wasdale, Lake District', *British Geological Survey Technical Report*, WG/88/7.

BLUNDELL, D. (1987): 'Carrock Mine, a short history', *The Mine Explorer: the Journal of the Cumbria Amenity Trust*, vol. 2, pp. 95-101.

BOTT, M. H. P. (1967): 'Geophysical investigations of the northern Pennine basement rocks', *Proceedings of the Yorkshire Geological Society*, vol. 36, pp. 139-68.

BOTT, M. H. P. (1974): 'The Geological interpretation of a gravity survey of the English Lake District and the Vale of Eden', *Journal of the Geological Society, London*, vol. 130, pp. 309-31.

BOTT, M. H. P. (1978): 'Deep Structure', pp. 25-40 in F. MOSELEY (ed.): *The Geology of the Lake District*, Yorkshire Geological Society, Occasional Publication, no. 3.

BOTT, M. H. P. and D. MASSON SMITH (1957): 'The Geological interpretation of a gravity survey of the Alston Block and Durham Coalfield' in *Quarterly Journal of the Geological Society of London*, vol. 113, pp. 93-117.

BOUCH, J. E., J. NADEN, T. J. SHEPHERD, J. A. MacKERVEY, B. YOUNG, A. J. BENHAM and H. J. SLOANE (2006): 'Direct evidence of fluid mixing in the formation of stratabound Pb-Zn-ba-F mineralisation in the Alston Block, North Pennine Orefield (England)', *Mineralium Deposita*, vol. 41, pp. 821-35.

BOWES, P. and A. ROBERTS (1983): 'Farmer-miners of the High Pennines', *Optima*, vol. 31, pp. 200-20.

BOZDAR, L. B. and B. A. KITCHENHAM (1972): 'Statistical appraisal of the occurrence of lead mines in the northern Pennines', *Transactions of the Institution of Mining and Metallurgy*, vol. 81, pp. B183-88.

BRADLEY, L. (1862): *An inquiry into the deposition of lead ore in the mineral veins of Swaledale, Yorkshire* (London: Stanford).

BRADSHAW, R. and F. C. PHILLIPS (1967): 'X-Ray studies on natural fabrics I: Growth fabrics in hematite kidney ore and in fibrous calcite', *Mineralogical Magazine*, vol. 36, pp. 70-77.

BRAITHWAITE, R. S. W. (1982): 'Wroewolfeite in Britain', *Mineralogical Record*, May-June 1982, pp. 167-74.

BRAITHWAITE, R. S. W., M. P. COOPER and A. D. HART (1989): 'Queitite, a mineral new to Britain from the Caldbeck Fells', *Mineralogical Magazine*, vol. 53, pp. 508-509.

BRAITHWAITE, R. S. W., T. B. GREENLAND and G. RYBACK (1963): 'Wulfenite from Poddy Gill, Caldbeck Fells', *Mineralogical Magazine*, vol. 33, p. 720.

BRAITHWAITE, R. S. W., R. P. H. LAMB and A. L. NATTRASS (2006): 'Heavy metals and their weathering products in residues from Castleside Smelting Mill, County Durham, including a new slag mineral, copper antimonate', *Journal of the Russell Society*, vol. 9, pp. 54-61.

BRAITHWAITE, R. S. W., R. P. H. LAMB and A. L. NATTRASS (2006): 'Secondary species associated with weathering of industrial residues at Tindale Fell Spelter works, Cumbria', *Journal of the Russell Society*, vol. 9, pp. 61-64.

BRAITHWAITE, R. S. W. and G. RYBACK (1988): 'Philipsburgite from the Caldbeck Fells and kipushite from Montana, and their infrared spectra', *Mineralogical Magazine*, vol. 52, pp. 529-33.

BRAITHWAITE, R. S. W. and G. RYBACK (1995): 'Beyerite, $(Ca,Pb)(BiO)_2(CO_3)_2$, a mineral new to the British Isles, from Cumbria, England', *Journal of the Russell Society*, vol. 6, pp. 51-52.

BRAMMALL, V. (1921): 'The mining, manufacture and uses of barytes in the neighbourhood of Appleby, Westmorland', *Transactions of the Institute of Mining Engineers*, vol. 61, pp. 42-45.

BRANDON, A., N. AITKENHEAD, R. G. CROFTS, R. A. ELLISON, D. J. EVANS and N. J. RILEY (1998): 'Geology of the Country around Lancaster. England and Wales', *Memoir of the British Geological Survey*.

BRANSTON, J. W. (1910): 'The minerals of Cumberland', *Transactions of the Carlisle Natural History Society*, vol. 2, pp. 14-29.

BREARS, P. (1989): *North Country Folk Art* (Donald Publication).

BREARS, P. (1992): 'Commercial Museums of 18th-century Cumbria, The Crosthwaite, Hutton and Todhunter Collections', *Journal of the History of Collections*, vol. 4, part 1, pp. 107-26.

BREITHAUPT, A. (1841): *Holoedrtes syntheticus oder Alstonit, Br. Vollstandiges Handbook der Mineralogy* (Dresden/Leipzig).

BRIDEN, J. C. and W. A. MORRIS (1973): 'Palaeomagnetic studies in the British Caledonides – III, Igneous Rocks of the Northern Lake District, England', *Journal of Geophysical Research*, vol. 34, 27-46.

BRIDGES, T. F. (1982): 'An occurrence of nickel minerals in the Hilton Mines, Scordale, Cumbria', *Journal of the Russell Society*, vol. 1, pp. 33-39.

BRIDGES, T. F. (1983): 'An occurrence of annabergite in Smallcleugh Mine, Nenthead, Cumbria', *Journal of the Russell Society*, vol. 1, p. 18.

BRIDGES, T. F. (1987): 'Serpierite and devilline from the Northern Pennine Orefield', *Proceedings of the Yorkshire Geological Society*, vol. 46, p. 169.

BRIDGES, T. F. and D. I. GREEN (2005): 'An update of the supergene mineralogy of Hilton Mine, Scordale, Cumbria', *Journal of the Russell Society*, vol. 8, pp. 108-10.

BRIDGES, T. F. and D. I. GREEN (2006): 'The first British occurrence of köttigite, from Hilton Mine, Scordale, Cumbria', *Journal of the Russell Society*, vol. 9, p. 3.

BRIDGES, T. F. and D. I. GREEN (2006): 'Baryte replacement by barium carbonate minerals', *Journal of the Russell Society*, vol. 9, pp. 73-82.

BRIDGES, T. F., D. I. GREEN, M. P. COOPER and N. THOMSON (2005): 'A review of the supergene mineralization at Silver Gill, Caldbeck Fells, Cumbria', *Journal of the Russell Society*, vol. 8, pp. 85-97.

BRIDGES, T. F., D. I. GREEN and M. S. RUMSEY (2005): 'A review of the occurrence of antlerite in the British Isles', *Journal of the Russell Society*, vol. 8, pp. 81-84.

BRIDGES, T. F., D. I. GREEN and M. S. RUMSEY (2006): 'A review of the mineralogy of Brae Fell Mine, Caldbeck Fells, Cumbria', *Journal of the Russell Society*, vol. 9, pp. 39-44.

BRIDGES, T. F. and H. WILKINSON (2003): 'Epimorphs of quartz after fluorite from the Rampgill, Coalcleugh and Barneycraig mine system, Nenthead, Cumbria, England', *Journal of the Russell Society*, vol. 8. pp. 38-39.

BRIDGES, T. F. and H. WILKINSON (2005): 'Epimorphs of quartz after fluorite from the Rampgill, Coalcleugh and Barneycraig mine system, Nenthead, Cumbria, England – an update', *Journal of the Russell Society*, vol. 8, p. 12.

BRIDGES, T. F. and B. YOUNG (1998): 'Supergene minerals of the Northern Pennine Orefield – a review', *Journal of the Russell Society*, vol. 7, 3-14.

BRISCOE, P. J. (2006): 'Unusual banded barite from Barras End Mine, Swaledale, North Yorkshire', *U.K. Journal of Mines and Minerals*, no. 27, pp. 11-14.

BRISTOW, H. W. (1861): *A glossary of mineralogy* (London: Longman, Green, Longman and Roberts).

BRITISH GEOLOGICAL SURVEY (BGS) (1988): *Geothermal energy in the UK: Review of the BGS Programme 1984-87. Investigation of Geothermal Potential of the UK* (Keyworth, Nottingham: BGS).

– BGS (1992): *Regional Geochemistry of the Lake District and Adjacent areas* (Keyworth, Nottingham: BGS).

– BGS (1996): *Regional Geochemistry of North-East England* (Keyworth, Nottingham: BGS).

BROCKBANK, W. (1869): 'The hematite iron ore deposits of Whitehaven: notes on the Aldby Limestone, Cleator', *Proceedings of the Manchester Literary and Philosophical Society*, vol. 8, pp. 51-55.

BROOKE, D., G. M. DAVIES, M. H. LONG and P. F. RYDER (1977): *Northern Caves, vol. 1 – The Northern Dales* (Clapham [via Lancaster]: Dalesman Books).

BROOKE, H. J. (1824): 'On baryto-calcite', *Annals of Philosophy*, vol. 8, pp. 114-16.

BROWN, G. C., R. A. IXER, J. A. PLANT and P. C. WEBB (1987): 'Geochemistry of granites beneath the north Pennines and their role in mineralisation', *Transactions of the Institution of Mining and Metallurgy*, vol. 96, pp. B65-76.

BROWN, M. J. (1983): 'Mineral investigations at Carrock Fell, Cumbria. Part 2 – Geochemical investigations', *Institute of Geological Sciences Mineral Reconnaissance Programme*, Report no. 60.

BUCKLEY, A. (1984): 'Anomalous uranium in superficial deposits overlying the Wasdale Head Granite, Cumbria', *Proceedings of the Yorkshire Geological Society*, vol. 44, pp. 431-42.

BULMAN, R. (2004): *Introduction to the Geology of Alston Moor* (Nenthead, Cumbria: North Pennines Heritage Trust).

BUNTING, R. (1994): 'Witherite from Scaleburn Vein, Nenthead, Cumbria, England', *Journal of the Russell Society*, vol. 5, pp. 118-19.

BURCHARD, U. and P. BODE (1986): *Mineral Museums of Europe* (Carson City, USA: Walnut Hill Publishing Company and Mineralogical Record).

BURGESS, I. C. and D. W. HOLLIDAY (1974): 'The Permo-Triassic Rocks of the Hilton Borehole, Westmorland – Part 1: Palaeogeography; Part 2 – Petrology of the Gypsum-Anhydrite Rocks', *Bulletin of the Geological Survey of Great Britain*, no. 46, pp. 1-34.

BURGESS, I. C. and D. W. HOLLIDAY (1979): 'Geology of the country around Brough-Under-Stainmore', *Memoir of the Geological Survey of Great Britain, England and Wales*, Sheet 31.

BURGESS, I. C. and A. J. WADGE (1974): *Classical areas of British geology: the geology of the Cross Fell Area: explanation of 1:25 000 Geological Special Sheet Comprising Parts of NY53, 62, 63, 64, and 1, 73* (London: HMSO, for Institute of Geological Sciences).

BURGESS, I. C. and A. J. WADGE (1974): *The geology of the Cross Fell Area* (London: HMSO, for Institute of Geological Sciences).

BURT, R. (1982): *A short history of British metal mining technology in the eighteenth and nineteenth centuries* (Aalst-Waarle: De Archaeologischee Pers Nederland).

BURT, R. (1982): *A short history of British ore preparation techniques in the eighteenth and nineteenth centuries* (Aalst-Waarle: De Archaeologischee Pers Nederland).

BURT, R. (1984): *The British lead mining industry* (Redruth: Dyllansow Truran).

BURT, R., P. WAITE and R. BURNLEY (1982): *The Cumberland mineral statistics: metalliferous and associated minerals 1845-1913*, Department of Economic History, University of Exeter with the Northern Mines Research Society and The Peak District Mines Historical Society.

CALVERT, J. (1854): *The gold rocks of Great Britain and Ireland* (London: Chapman and Hall).

CAMERON, A. (2000): *Mining Heritage Lakelands – The Last 500 Years*, Cumbria Amenity Trust Mining History Society.

CANN, J. R. and D. A. BANKS (2001): 'Constraints on the genesis of the mineralisation of the Alston Block, Northern Pennines', *Proceedings of the Yorkshire Geological Society*, vol. 53, pp. 187-96.

CANNEL, A. E. and M. CANNELL (1968): 'The Iron Ore Mines of Tongue Gill' (pp. 65-66); 'Hartsop Hall Mines' (pp. 67-68); 'Lowthwaite Mines, Cumberland (pp. 68-69); and 'Grisedale Lead Mines, Patterdale', pp. 69-70, in *Memoirs of the Northern Cavern and Mines Society*, vol. 1(7).

CARRUTHERS, R. G., G. A. BURNETT and W. ANDERSON (1932): 'The geology of the Cheviot Hills', *Memoir of the Geological Survey of Great Britain*.

CARRUTHERS, R. G., C. H. DINHAM, G. BURNETT, J. MADEN (1927): 'Geology of Belford, Holy Island and Farne Islands', *Memoir of the Geological Survey of Great Britain*.

CARRUTHERS, R. G. and R. W. POCOCK (1922): 'Fluorspar' (3rd edition), Special Reports on the Mineral Resources of Great Britain, *Memoir of the Geological Survey of Great Britain*, vol. 4.

CARRUTHERS, R. G. and A. STRAHAN (1923): 'Lead and zinc ores of Durham, Yorkshire and Derbyshire, with notes on the Isle of Man', Special Reports on the Mineral Resources of Great Britain, *Memoir of the Geological Survey of Great Britain*, vol. 24.

CHADWICK, R. A., D. W. HOLLIDAY, S. HOLLOWAY and A. G. HULBERT (1995): 'The Northumberland-Solway basin and adjacent areas', *Subsurface Memoir of the British Geological Survey*.

CHADWICK, R. A., D. I. JACKSON, R. P. BARNES, G. S. KIMBELL, H. JOHNSON, R. C. CHIVERRELL, G. S. P. THOMAS, N. S. JONES, N. J. RILEY, E. A. PICKETT, B. YOUNG, D. W. HOLLIDAY, D. F. BALL, S. G. MOLYNEUX, D. LONG, G. M. POWER and D. H. ROBERTS (2001): 'Geology of the Isle of Man and its off-shore area', *British Geological Survey Research Report*, RR/01/06.

CHAMBERS, B. (ed.) (1992): *Men, mines and minerals of the North Pennines* (Killhope, Co. Durham: Friends of Killhope).

CHAMBERS, B. (ed.) (1997): *Out of the Pennines* (Killhope, Co. Durham: Friends of Killhope).

CHAMBERS, B. (ed.) (2002): *Friends of the northern lead dales* (Killhope, Co. Durham: Friends of Killhope).

CHISHOLM, J. E., G. C. JONES and O. W. PURVIS (1987): 'Hydrated copper oxalate, moolooite, in lichens', *Mineralogical Magazine*, vol. 51, pp. 715-18.

CLAPHAM, R. C. and J. DAGLISH (1884): 'On minerals and salts found in coal pits', *Transactions of the North of England Institute of Mining Engineers*, vol. 13, pp. 219-26.

CLARINGBULL, G. F. (1951): 'New occurrences of duftite', *Mineralogical Magazine*, vol. 29, pp. 609-14.

CLARIS, P., J. QUATERMAINE and A. R. WOOLEY (1989): 'The Neolithic axe quarries and axe factory sites of Great Langdale and Scafell Pike: A new field survey', *Proceedings of the Prehistoric Society*, vol. 55, pp. 1-25.

CLEAL, C. J. and B. A. THOMAS (1996): *British Upper Carboniferous Stratigraphy* (London: Chapman and Hall).

CLOUGH, R. T. (1962): *The lead smelting mills of the Yorkshire Dales* (Leeds: Clough).

CLOUGH, R. T. (1967): 'Catalogue of a collection of Yorkshire mine plans relative to the Greenhow area', *Memoirs of the Northern Cavern and Mines Research Society*, pp. 11-15.

CLOUGH, R. T. (1980): *Lead smelting mills of the Yorkshire dales and Northern Pennines* (Keighley: Clough).

CLOWES, F. (1889): 'Deposits of barium sulphate from mine water', *Proceedings of the Royal Society, London*, vol. 46, pp. 368-69.

COLLINGWOOD, W. G. (1910): 'Germans at Coniston in the seventeenth century', *Transactions of the Cumberland and Westmorland Antiquarian and Archaeological Society* (new series), vol. 10, pp. 369-94.

COLLINGWOOD, W. G. (1912): 'Elizabethan Keswick: extracts from the original accounts books 1564-1577, of the German miners, in the archives of Augsburg', *Transactions of the Cumberland and Westmorland Antiquarian and Archaeological Society* (tract series), no. 8.

COLLINGWOOD, W. G. (1928): 'The Keswick and Coniston Mines in 1600 and later', *Transactions of the Cumberland and Westmorland Antiquarian and Archaeological Society*, vol. 28, pp. 1-32.

COOMER, P. G. and T. D. FORD (1975): 'Lead and sulphur isotope ratios of some galena specimens from the Pennines and north Midlands', *Mercian Geologist*, vol. 5, pp. 291-304.

COOPER, A. H. and S. G. MOLYNEUX (1990): 'The age and correlation of the Skiddaw Group (early Ordovician) sediments in the Cross Fell inlier (northern England)', *Geological Magazine*, vol. 127, pp. 147-57.

COOPER, D. C., D. G. CAMERON, B. YOUNG, J. D. CORNWELL and D. J. BLAND (1991): 'Mineral exploration in the Cockermouth area, Cumbria – Part 1: regional surveys', *British Geological Survey*, Technical Report no. WF/91/4 (*BGS Mineral Reconnaissance Programme*, Report 118).

COOPER, D. C., D. G. CAMERON, B. YOUNG, B. C. CHACKSFIELD and J. D. CORNWELL (1992): 'Mineral exploration in the Cockermouth area, Cumbria – Part 2: follow up surveys', *British Geological Survey*, Technical Report no. WF/92/3 (*BGS Mineral Reconnaissance Programme*, Report 122).

COOPER, D. C., M. K. LEE, N. J. FORTEY, A. H. COOPER, C. C. RUNDLE, B. C. WEBB and P. M. ALLEN (1988): 'The Crummock Water aureole: a zone of metasomatism and source of ore metals in the English Lake District', *Journal of the Geological Society, London*, vol. 145, pp. 523-40.

COOPER, E. (1960): *Men of Swaledale: an account of Yorkshire farmers and miners* (Clapham [via Lancaster]: Dalesman).

COOPER, M. P. (2006): 'Robbing the Sparry Garniture', *A 200 year History of British Mineral Dealers. 1750-1950* (Tuscon, Arizona: Mineralogical Record Inc.).

COOPER, M. P., D. I. GREEN and R. S. W. BRAITHWAITE (1988): 'The occurrence of mattheddleite in the Caldbeck Fells, Cumbria: a preliminary note', *U.K. Journal of Mines and Minerals*, no. 5, p. 21.

COOPER, M. P. and C. J. STANLEY (1990): *Minerals of the English Lake District – Caldbeck Fells* (London: British Museum [Natural History]).

CORNWELL, J. D., D. J. PATRICK and J. M. HUDSON (1978): 'Geophysical investigations along parts of the Dent and Augill faults', *Institute of Geological Sciences Mineral Reconnaissance Report*, no. 24.

CRABTREE, P. and R. FOSTER (1963): 'Sir Francis Mine', *Cave Science*, vol. 5, pp. 1-24.

CRABTREE, P. and R. FOSTER (1965): 'The Kisdon Mining Company', *Bulletin of the Peak District Mines Historical Society*, vol. 2, pp. 303-306; vol. 3, pp. 63-67, 119-24.

CREANEY, S. (1980): 'Petrographic texture and vitrinite reflectance variation on the Alston Block', *Proceedings of the Yorkshire Geological Society*, vol. 42, pp. 553-80.

CREANEY, S. (1982): 'Vitrinite reflectance determinations from the Beckermonds Scar and Raydale boreholes, Yorkshire', *Proceedings of the Yorkshire Geological Society*, vol. 44, pp. 99-102.

CREANEY, S., D. M. ALLCHURCH and J. M. JONES (1979): 'Vitrinite reflectance variation in northern England', *C. R. 9e Congr. Int. Stratigr, Geol. Carbonif. Abs.*, pp. 46-47.

CRIDDLE, A. J. and C. J. STANLEY (1979): 'New data on wittichenite', *Mineralogical Magazine*, vol. 43, 109-13.

CRITCHLEY, M. F. (1984): 'The history and working of the Nenthead mines', *Bulletin of the Peak District Mines Historical Society*, vol. 9, pp. 1-50.

CROSBY, J. (1989): *Weardale in Old Photographs* (Alan Sutton).

CROSSLEY, C. and K. PATRICK (undated): *Blanchland's lead mining heritage* (Northumberland County Council).

CROSTHWAITE, J. F. (1876): 'Peter Crosthwaite, The Founder of Crosthwaite's Museum, Keswick', *Transactions of the Cumberland and Westmoreland Antiquarian and Archaeological Society*, vol. 1, pp. 151-64.

CROWLEY, S. F., S. H. BOTTRELL, B. McCARTHY, J. WARD AND B. YOUNG (1997): 'δ^{34}S of Lower Carboniferous anhydrite, Cumbria and its implications for barite mineralization in the northern Pennines', *Journal of the Geological Society of London*, vol. 154, pp. 597-600.

CUMBRIA AMENITY TRUST MINING HISTORY SOCIETY (1992): *Beneath the Lakeland Fells* (Ulverston, Cumbria: Red Earth Publications).

CURTO, C. (2006): 'Colour in fluorite', *Mineral Up*, vol. 1, pp. 32-41.

DAGGER, G. W. (1977): 'Controls of copper mineralization at Coniston, English Lake District', *Geological Magazine*, vol. 114, pp. 195-202.

DAKYNS, J. R., R. RUSSELL, C. T. CLOUGH and A. STRAHAN (1891): 'The Geology of the Country around Mallerstang, with parts of Wensleydale and Arkendale', *Memoir of the Geological Survey of Great Britain*.

DAKYNS, J. R., R. H. TIDDEMAN and J. G. GOODCHILD (1897): 'Geology of the country between Appleby, Ullswater and Haweswater', *Memoir of the Geological Survey of Great Britain*.

DAKYNS, J. R., R. H. TIDDEMAN, W. GUNN and A. STRAHAN (1890): 'The geology of the country around Ingleborough, with parts of Wensleydale and Wharfedale', *Memoir of the Geological Survey of Great Britain*.

DAVIDSON, W. F. (1957): 'Mines and minerals', pp. 66-63 in 'The changing scene', *Joint Transactions of the Eden Field Club, Penrith and District Natural History Society and Kendal Natural History Society*, no. 1.

DAVIDSON, W. F. and N. THOMSON (1948): 'Some notes on the minerals of Cumberland and Westmorland', *The North Western Naturalist*, vol. 23, pp. 136-54.

DAVIES, W. J. K. (1981): *The Ravenglass and Eskdale Railway* (2nd edition) (Newton Abbot: David and Charles).

DAVISON, W., J. HILTON, J. P. LISHMAN and W. PENNINGTON (1985): 'Contemporary lake transport processes determined from sedimentary records of copper mining activity', *Environmental Science and Technology*, vol. 19, pp. 356-60.

DAWSON, E. W. O. (1947): 'War-time treatment of lead-zinc dumps situated at nenthead, Cumberland', *Transactions of the Institution of Mining and Metallurgy*, vol. 56, pp. 587-606.

DAWSON, J. and R. K. HARRISON (1966): 'Uraninite in the Grainsgill Greisen, Cumberland', *Bulletin of the Geological Survey of Great Britain*, no. 25, p. 91.

DAY, F. H. (1928): 'Some notes on the minerals of Caldbeck Fells', *Transactions of the Carlisle Natural History Society*, vol. 4, pp. 66-79.

DAY, J. B. W. (1970): 'Geology of the country around Bewcastle', *Memoir of the Geological Survey of Great Britain*.

DAYSH, G. H. J. and E. M. WATSON (1951): *Cumberland with special reference to the West Cumberland development area. A survey of industrial facilities* (Whitehaven: the Cumberland Development Council Ltd and the West Cumberland industrial Development Co. Ltd).

DEANS, T. (1950): 'The Kupferschiefer and associated lead-zinc mineralisation in the Permian of Silesia, Germany and England', *Report of the 18th international Geological Congress* (London), part 7, pp. 340-52.

DEANS, T. (1951): 'Notes on the copper deposits of Middleton Tyas and Richmond', Abstract in *Mineralogical Society Notice*, no. 74.

DEARMAN, W. R. and J. M. JONES (1967): 'Millerite from Boldon Colliery', *Transactions of the Natural History Society of Northumberland*, vol. 16, pp. 193-96.

DEWEY, H. (1920): 'Arsenic and antimony ores', Special Reports on the Mineral Resources of Great Britain, *Memoir of the Geological Survey of Great Britain*, vol. 15.

DEWEY, H. and H. C. DINES (1923): 'Tungsten and manganese ores', Special Reports on the Mineral Resources of Great Britain, *Memoir of the Geological Survey of Great Britain*, vol. 1.

DEWEY, H. and T. EASTWOOD (1925): 'Copper ores of the Midlands, the Lake District and the Isle of Man', Special Reports on the Mineral Resources of Great Britain, *Memoir of the Geological Survey of Great Britain*, vol. 30.

DICKENSON, J. (1902): 'Lead mining districts of the North of England and Derbyshire', *Transactions of the Manchester Geological Society*, vol. 27, pp. 218-65.

DICKINSON, J. M. (1964): 'Some notes on the lead mines of Greenhow hill', *Transactions of the Northern Cavern and Mines Research Society*, vol. 1, pp. 34-47.

DICKINSON, J. M. (1964): 'The Appletreewick Lead Mining Company 1870-72', *Memoirs of the Northern Cavern and Mines Research Society*, vol. 1(1), pp. 1-7; vol. 1(2), pp 1-6; vol. 1(3), pp. 1-4.

DICKINSON, J. M. (1965): 'The Dales Chemical Company', *Memoirs of the Northern Cavern and Mines Research Society*, vol. 1(4), pp. 1-5.

DICKINSON, J. M. (1966): 'The Grimwith Lead Mining Company Limited', *Memoirs of the Northern Cavern and Mines Research Society*, vol. 1(5), pp. 1-21.

DICKINSON, J. M. (1967): 'Burhill Mine', *Memoirs of the Northern Cavern and Mines Research Society*, vol. 1(6), pp. 1-9.

DICKINSON, J. M. (1968): 'The Rimmington Lead and Silver Mines', *Memoirs of the Northern Cavern and Mines Research Society*, vol. 1(7), pp. 21-22.

DICKINSON, J. M. (1969): 'The Greenhow Lead Mining Field', *Memoirs of the Northern Cavern and Mines Research Society*, vol. 1(8), pp. 25-31.

DICKINSON, J. M. (1970): 'The Greenhow Lead Mining Field (A historical survey)', *Northern Cavern and Mine Research Society*, Individual Survey Series, no. 4.

DICKINSON, J. M. (1971): 'Dry Gill Mill', *Transactions of the Northern Cavern and Mines Research Society*, vol. 2, pp. 31-36.

DICKINSON, J. M. (1972): 'The mineral veins of Swinden Knoll', *Transactions of the Northern Cavern and Mines Research Society*, vol. 2(2), pp. 98-100.

DICKINSON, J. M. (1973): 'Old Providence Lead Mine', *Memoirs of the Northern Cavern and Mines Research Society*, vol. 2(3), pp. 101-104.

DICKINSON, J. M. and M. C. GILL (1983): 'The Greenhow Lead Mining Field – A Historical Survey', *British Mining*, no. 21, *Monograph of the Northern Mine Research Society*.

DICKINSON, J. M. and M. C. MAY (1968): 'Wharfedale Mines', *Memoirs of the Northern Cavern and Mines Research Society*, pp. 9-12.

DINES, H. G. (1922): 'Barytes and Witherite' (3rd edition), Special Report on the Mineral Resources of Great Britain, *Memoir of the Geological Survey*, vol. 2.

DIXON, E. E. L. (1928): 'The Origin of Cumberland Haematites', *Summary of Progress of the Geological Survey for 1927*, pp. 23-32.

DIXON, E. E. L., J. MADEN, F. M. TROTTER, S. E. HOLLINGWORTH and L. H. TONKS (1926): 'The geology of the Carlisle, Longtown and Silloth district', *Memoir of the Geological Survey of Great Britain*.

DODD, M. (ed.) (1982): *Lakeland rocks and landscape. A field guide* (Maryport, Cumbria: Cumberland Geological Society and Ellenbank Press).

DONALD, M. B. (1994): *Elizabethan copper* (Ulverston, Cumbria: Red Earth Publications).

DUBOIS, R. L. (1962): 'Magnetic characteristics of a massive hematite body', *Journal of Geophysical Research*, vol. 67, pp. 2887-93.

DUNHAM, A. C. (1970): 'Whin sills and dykes', in 'Geology of Durham County' (pp. 92-100), *Transactions of the Natural History Society of Northumberland, Durham and Newcastle Upon Tyne*, vol. 41.

DUNHAM, K. C. (1931): 'Mineral deposits of the north Pennines', *Proceedings of the Geologists' Association*, vol. 42, pp. 274-81.

DUNHAM, K. C. (1934): 'The genesis of the north Pennine ore deposits', *Quarterly Journal of the Geological Society of London*, vol. 90, pp. 689-720.

DUNHAM, K. C. (1937): 'The paragenesis and colour of fluorite in the English Pennines', *American Mineralogist*, vol. 22, pp. 468-79.

DUNHAM, K. C. (1941): 'Iron ore deposits of the northern Pennines', *Geological Survey Wartime Pamphlet*, no. 14.

DUNHAM, K. C. (1944): 'The production of galena and associated minerals in the northern Pennines; with comparative statistics for Great Britain', *Transactions of the Institution of Mining and Metallurgy*, vol. 53, pp. 181-252.

DUNHAM, K. C. (1947): 'Excursion to Durham and the Northern Pennines', *Mineralogical Magazine*, vol. 28, pp. 45-51.

DUNHAM, K. C. (1948): 'Geology of the Northern Pennine Orefield' (1st edition), Vol. 1 – Tyne to Stainmore, *Memoir of the Geological Survey of Great Britain*.

DUNHAM, K. C. (1952): 'Age-relations of the Epigenetic Mineral Deposits of Britain', *Transactions of the Geological Society of Glasgow*, vol. 21, pp. 395-429.

DUNHAM, K. C. (1952): 'Fluorspar' (4th edition), Special Reports on the Mineral Resources of Great Britain, *Memoir of the Geological Survey of Great Britain*, vol. 4.

DUNHAM, K. C. (1959): 'Epigenetic mineralization in Yorkshire', *Proceedings of the Yorkshire Geological Society*, vol. 32, pp. 1-29.

DUNHAM, K. C. (1959): 'Non-ferrous mining potentialities of the Northern Pennines', pp. 115-48 in *The future of non-ferrous mining in Great Britain and Ireland* (London: Institution of Mining and Metallurgy).

DUNHAM, K. C. (1964): 'Neptunist concepts in ore genesis', *Economic Geology*, vol. 59, pp. 1-21.

DUNHAM, K. C. (1966): 'Role of juvenile solutions, connate waters and evaporitic brines in the genesis of lead-zinc-fluorine-barium deposits', *Transactions of the Institution of Mining and Metallurgy*, vol. 75, pp. B226-29; discussion, pp. 300-305.

DUNHAM, K. C. (1967): 'Mineralization in relation to pre-Carboniferous basement rocks, northern England', *Proceedings of the Yorkshire Geological Society*, vol. 36, pp. 195-201.

DUNHAM, K. C. (1967): 'Veins, flats and pipes in the Carboniferous of the English Pennines', *Economic Geology Monograph*, no. 3, pp. 201-207.

DUNHAM, K. C. (1970): 'Mineralization', pp. 124-31 in 'Geology of Durham County', *Transactions of the Natural History Society of Northumberland, Durham and Newcastle Upon Tyne*, vol. 41.

DUNHAM, K. C. (1970): 'Mineralization by deep formation waters: a review', *Transactions of the Institution of Mining and Metallurgy*, vol. 79, pp. B127-36.

DUNHAM, K. C. (1974): 'Epigenetic Minerals', pp. 294-308 in D. H. RAYNER and J. E. HEMINGWAY (eds): *Geology and Mineral Resources of Yorkshire*, Yorkshire Geological Society.

DUNHAM, K. C. (1974): 'Granite beneath the Pennines in north Yorkshire', *Proceedings of the Yorkshire Geological Society*, vol. 40, pp. 191-94.

DUNHAM, K. C. (1978): 'Weardale minerals', *Journal of the Weardale Field Studies Society*, vol. 1, pp. 5-9.

DUNHAM, K. C. (1983): 'Ore genesis in the English Pennines: a fluoritic subtype', *Proceedings of the International Conference on Mississippi Valley Type Lead-Zinc Deposits, Rolla, Missouri*, pp. 86-112.

DUNHAM, K. C. (1984): 'Genesis of the Cumbria hematite deposits', *Proceedings of the Yorkshire Geological Society*, vol. 45, p. 130.

DUNHAM, K. C. (1988): 'Pennine mineralisation in depth', *Proceedings of the Yorkshire Geological Society*, vol. 47, pp. 1-12.

DUNHAM, K. C. (1990): 'Geology of the Northern Pennine Orefield' (2nd edition), Vol. 1 – Tyne To Stainmore, *Economic Memoir of the British Geological Survey, England and Wales*.

DUNHAM, K. C., K. E. BEER, R. A. ELLIS, M. J. GALLAGHER, M. J. C. NUTT, and B. C. WEBB (1979): 'United Kingdom', pp. 263-317 in S. H. U. BOWIE, A. KVALHEIM and H. W. HASLAM (eds): *Mineral deposits of Europe*, 1 (London: Institute of Mining and Metallurgy Society).

DUNHAM, K. C., M. P. BOTT, G. A. L. JOHNSON and B. L. HODGE (1961): 'Granite beneath the Northern Pennines', *Nature* (London), vol. 190, pp. 899-900.

DUNHAM, K. C. and H. G. DINES (1945): 'Barium minerals in England and Wales', *Geological Survey Wartime Pamphlet*.

DUNHAM, K. C., A. C. DUNHAM, B. L. HODGE and G. A. L. JOHNSON (1965): 'Granite beneath Viséan sediments with mineralisation at Rookhope, north Pennines', *Quarterly Journal of the Geological Society of London*, vol. 121, pp. 383-417.

DUNHAM, K. C., F. J. FITCH, P. R. INESON, J. A. MILLER and J. G. MITCHELL (1968): 'The geochronological significance of argon-40/argon-39 age determinations on white whin from the northern Pennine orefield', *Proceedings of the Royal Society, London*, vol. 307, pp. 251-66.

DUNHAM, K. C. and W. C. C. ROSE (1941): 'Geology of the iron ore field of south Cumberland and Furness', *Wartime Pamphlet of the Geological Survey of Great Britain*, no. 16.

DUNHAM, K. C. and C. J. STUBBLEFIELD (1945): 'The stratigraphy, structure and mineralisation of the Greenhow mining area', *Quarterly Journal of the Geological Society of London*, vol. 100, pp. 209-68.

DUNHAM, A. C. and V. E. H. TRASSER-KNIG (1982): 'Late Carboniferous intrusions of northern Britain', in D. S. SUTHERLAND (ed.): *Igneous rocks of the British Isles* (New York: John Wiley).

DUNHAM, K. C. and A. A. WILSON (1985): 'Geology of the Northern Pennine Orefield', Vol. 2 – Stainmore to Craven, *Economic Memoir of the British Geological Survey*.

DUNHAM, K. C., B. YOUNG, G. A. L. JOHNSON, T. B. COLMAN and R. FOSSETT (2001): 'Rich silver-bearing ores in the Northern Pennines?', *Proceedings of the Yorkshire Geological Society*, vol. 53, pp. 207-12.

DURHAM COUNTY COUNCIL (no date): *Lead and life at Killhope*, Durham County Council.

EASTWOOD, T. (1921): 'The lead and zinc ores of the Lake District', Special Reports on the Mineral Resources of Great Britain, *Memoir of the Geological Survey of Great Britain*, vol. 22.

EASTWOOD, T. (1930): 'The geology of the Maryport District', *Memoir of the Geological Survey of Great Britain*.

EASTWOOD, T. (1959): 'The Lake District Mining Field', pp. 149-74 in *The future of non-ferrous mining in Great Britain and Ireland* (London: Institution of Mining and Metallurgy).

EASTWOOD, T., E. E. L. DIXON, S. E. HOLLINGWORTH and B. SMITH (1931): 'The geology of the Whitehaven and Workington district', *Memoir of the Geological Survey of Great Britain, England and Wales*.

EASTWOOD, T., S. E. HOLLINGWORTH, W. C. C. ROSE and F. M. TROTTER (1968): 'Geology of the country around Cockermouth and Caldbeck', *Memoir of the Geological Survey of Great Britain, England and Wales*.

EDDY, J. R. (1883): 'On the lead veins in the neighbourhood of Skipton', *Proceedings of the Yorkshire Geological and Polytechnic Society*, vol. 8, pp. 63-69.

EDDY, S. (1844): 'On the geology of the Grassington mines, near Skipton, Yorkshire', *Transactions of the Royal Geological Society of Cornwall*, vol. 6, pp. 186-89.

EDDY, S. (1845): 'An account of the Grassington lead mines, illustrating a model of a mine', *Report of the British Association for the Advancement of Science for 1844*, pp. 52-53.

EDDY, S. (1859): 'On the lead mining districts of Yorkshire', *Proceedings of the Yorkshire Geological and Polytechnic Society*, vol. 3, pp. 657-63.

EDDY, S. (1859): 'On the lead mining districts of Yorkshire', *Report of the British Association for the Advancement of Science for 1858*, pp. 167-74.

EGGLESTONE, W. M. (1882): *Stanhope and its Neighbourhood* (Stanhope).

EGGLESTONE, W. W. (1908): 'The occurrence and commercial

uses of fluorspar', *Transactions of the North of England Institute of Mining Engineers*, vol. 35, pp. 236-68.

EL SHAZLY, E. M., J. S. WEBB and D. WILLIAMS (1957): 'Trace elements in sphalerite, galena and associated minerals from the British Isles', *Transactions of the Institution of Mining and Metallurgy*, 66, pp. 241-71.

ELLIS, R. (ed.) (1851): *Official descriptive and illustrated catalogue of the Great Exhibition of the Works of Industry of all Nations* (London: Spicer Bros).

EMBREY, P. G. (1977): 'Fourth supplementary list of British minerals', in P. R. GREG and W. G. LETTSOM (1858): *Manual of the Mineralogy of Great Britain and Ireland* (Broadstairs, Kent: Lapidary Publications, reprint 1977).

EMBREY, P. G. and R. F. SYMES (1987): *Minerals of Cornwall and Devon* (London: British Museum [Natural History]).

EMELEUS, H. J. (1970): 'Tertiary igneous intrusions', pp. 101-102 in 'Geology of Durham County', *Transactions of the Natural History Society of Northumberland, Durham and Newcastle Upon Tyne*, vol. 41.

EVANS, A. D., D. J. PATRICK, A. J. WADGE and J. M. HUDSON (1983): 'Geophysical investigations in Swaledale', *Institute of Geological Sciences Mineral Reconnaissance Report*, no. 65.

EVANS, A. M. (1986): 'Comments on Sir Kingsley Dunham's Paper: Age and Origin of the Cumbrian Hematite', *Proceedings of the Cumberland Geological Society*, vol. 4, pp. 405-406.

EVANS, A. M. and A. EL-NIKHELY (1982): 'Some palaeomagnetic dates from the West Cumbrian hematite deposits, England', *Transactions of the Institution of Mining and Metallurgy* (Section B: Applied Earth Science), vol. 91, B41-B43.

EWART, A. (1962): 'Hydrothermal alteration in the Carrock Fell area, Cumberland, England', *Geological Magazine*, vol. 99, pp. 1-8.

EYLES, V. A. (1965): 'John Woodward FRS (1665-1728), *Nature*, vol. 2016, pp. 868-70.

FAIRBAIRN, R. A. (1993): 'The mines of Alston Moor', *Monograph of the Northern Mines Research Society*.

FAIRBAIRN, R. A. (1996): 'Weardale Mines', *Monograph of the Northern Mine Research Society*, no. 56.

FAIRBAIRN, R. A. (ed.) (2003): *Fluorspar in the North Pennines* (Killhope, Co. Durham: Friends of Killhope).

FAIRBAIRN, R. A. (2004): 'The Mines of Upper Teesdale', *British Mining no. 77, Monograph of the Northern Mines Research Society*.

FEATHER, S. W. (1966): 'Saline Spring, Brandlehow Mine, near Keswick, *Memoirs of the Northern Cavern and Mines Research Society*, vol. 1(5), pp. 53-54.

FIELDHOUSE, R. and B. JENNINGS (1978): *A history of Richmond and Swaledale* (London: Phiullimore).

FINLAYSON, A. M. (1910): 'The ore-bearing Pegmatites of Carrock Fell, and the genetic significance of Tungsten-Ores', *Geological Magazine*, vol. 7, pp. 19-28.

FIRMAN, R. J. (1953): 'On the occurrence of nacrite at Shap, Westmorland', *Mineralogical Magazine*, vol. 30, pp. 199-200.

FIRMAN, R. J. (1957): 'Fissure metasomatism in volcanic rocks adjacent to the Shap granite, Westmorland', *Quarterly Journal of the Geological Society of London*, vol. 113, pp. 205-22.

FIRMAN, R. J. (1978): 'Epigenetic mineralisation', pp. 226-41 in F. MOSELEY (ed.): *The Geology of the Lake District*, Yorkshire Geological Society, Occasional Publication, no. 3.

FIRMAN, R. J. (1978): 'Intrusions', pp. 146-63 in F. MOSELEY (ed.): *The Geology of the Lake District*, Yorkshire Geological Society, Occasional Publication, no. 3.

FIRMAN, R. J. and M. K. LEE (1986): 'The age and structure of the concealed English Lake District batholith and its probable influence on subsequent sedimentation, tectonics and mineralisation', in R. W. NESBITT and J. NICHOL (eds): *Geology in the Real World – The Kingsley Dunham Volume*, Institution of Mining and Metallurgy, pp. 117-27.

FISHER, J. (2003): 'The Rogerley Mine, Weardale, Co. Durham', *U.K. Journal of Mines and Minerals*, no. 23, pp. 9-20.

FISHER, J. (2004): 'Fluorite from the Northern Pennines Orefield, England', *Rocks and Minerals*, no. 79, pp. 378-98.

FISHER, J. (2006): 'Classic 19th-century fluorite specimens from Weardale, Durham: a mineralogical mystery', *U.K. Journal of Mines and Minerals*, no. 27, pp. 25-28.

FISHER, J. and L. GREENBANK (2000): 'The Rogerley Mine, Weardale, Co. Durham, England', *Rocks and Minerals*, no. 75, pp. 54-61.

FITCH, F. J. and J. A. MILLER (1965): 'Age of the Weardale Granite', in *Nature* (London), vol. 208, pp. 743-45.

FITTON, J. G. (1972): 'The genetic significance of almandine-pyrope phenocrysts in the calc-alkaline Borrowdale Volcanic Group, northern England', *Contributions to Mineralogy and Petrology*, vol. 36, pp. 231-48.

FITTON, J. G., M. F. THIRLWALL and D. J. HUGHES (1982): 'Volcanism in the Caledonian orogenic belt of Britain', pp. 611-36 in R. S. THORPE (ed.): *Andesites* (New York: Wiley).

FORBES, I. (undated): *Secret worlds: Spar boxes of the North Pennines* (Killhope, Co. Durham: Lead Mining Museum).

FORBES, I. (2003): 'The twentieth century fluorspar mining industry', pp. 1-3 in R. A. FAIRBAIRN (ed.): *Fluorspar in the Pennines* (Killhope, Co. Durham: Friends of Killhope).

FORBES, I., B. YOUNG, C. CROSSLEY and L. HEHIR (2003): *The lead mining landscape of the North Pennines Area of Outstanding Natural Beauty* (Durham County Council).

FORSTER, W. (1809): *A treatise on a section of the strata from Newcastle-Upon-Tyne to the mountain of Cross Fell in Cumberland, with remarks on mineral veins in general* (1st edition) (Alston: Forster).

FORSTER, W. (1821): *A treatise on a section of the strata from Newcastle-Upon-Tyne to the mountain of Cross Fell in Cumberland, with remarks on mineral veins in general* (2nd edition) (Alston: Forster).

FORSTER, W. (1883): *A treatise on a section of the strata from*

Newcastle-Upon-Tyne to the mountain of Cross Fell in Cumberland, with remarks on mineral veins in general (3rd edition, revised by W. NALL) (Newcastle upon Tyne: Andrew Reid).

FORTEY, N. J. (1978): 'Mineral parageneses of the Harding vein tungsten deposit, Carrock Fell mine, Cumbria', *Mineralogy Unit Report no. 228, Institute of Geological Sciences*.

FORTEY, N. J. (1980): 'Hydrothermal mineralization associated with minor late Caledonian intrusions in northern Britain: preliminary comments', *Transactions of the Institution of Mining and Metallurgy*, vol. 89, pp. B173-76.

FORTEY, N. J. and D. J. BLAND (1979): 'Heavy mineral dispersion profiles in altered granite at Carrock Fell Mine, Cumbria', *Applied Mineralogy Unit Report no. 245, Institute of Geological Sciences*.

FORTEY, N. J., D. J. BLAND, J. NADEN, B. YOUNG, D. MILLWARD and M. PÉREZ-ALVAREZ (1994): 'Zonation patterns in sphalerite crystals from the Caldbeck Fells mining district, English Lake District' (conference abstract), *Transactions of the Institution of Mining and Metallurgy*, vol. 103, pp. B210-11.

FORTEY, R. A., D. A. T. HARPER, J. K. INGHAM, A. W. OWEN, M. A. PARKES, A. W. A. RUSHTON and N. H. WOODCOCK (2000): 'A revised correlation of Ordovician rocks in the British Isles', *Geological Society, London*, Special Report no. 24, 83 pp.

FORTEY, N. J., J. D. INGHAM, B. R. H. SKILTON, B. YOUNG and T. J. SHEPHERD (1984): 'Antimony mineralisation at Wet Swine Gill, Caldbeck Fells, Cumbria', *Proceedings of the Yorkshire Geological Society*, vol. 45, pp. 59-65.

FORTEY, N. J. and P. H. A. NANCARROW (1990): 'Chromium minerals in microdiorite from the English Lake District', *Journal of the Russell Society*, vol. 3, pp. 15-22.

FÖRTSCH, E. B. (1967): '"Plumbogummite" from Roughten Gill, Cumberland', *Mineralogical Magazine*, vol. 36, pp. 59-65.

FOSTER-SMITH, J. R. (1968): 'Notes on the lead mines on the east side of Wharfedale between Buckden and Kettlewell', *Memoirs of the Northern Cavern and Mines Research Society*, vol. 1(7), pp. 14-21.

FOWLER, A. (1926): 'The geology of Berwick-Upon-Tweed, Norham and Scremerston', *Memoir of the Geological Survey of Great Britain*.

FOWLER, A. (1936): 'The geology of the country around Rothbury, Amble and Ashington', *Memoir of the Geological Survey of Great Britain*.

FOWLER, A. (1943): 'On Fluorite and other minerals in Lower Permian Rocks of South Durham', *Geological Magazine*, vol. 80, pp. 41-51.

FOWLER, A. (1956): 'Minerals in the Permian and Trias of north-east England', *Proceedings of the Geologists' Association*, vol. 67, pp. 251-65.

FROST, D. V. and D. W. HOLLIDAY (1980): 'Geology of the country around Bellingham', *Memoir of the Geological Survey of Great Britain*.

GEDDES, R. S. (1975): *Burlington Blue-Grey. A history of the Slate Quarries, Kirkby-in-Furness* (Kirkby-in-Furness: Geddes).

GILL, M. C. (1974): 'Buckden Gavel Lead Mine', *Memoirs of the Northern Cavern and Mines Research Society*, vol. 2(4), pp. 179-82.

GILL, M. C. (1974): 'Lolly Scar and Blayshaw Gill Lead Mines', *Memoirs of the Northern Cavern and Mines Research Society*, vol. 2(4), pp. 203-204.

GILL, M. C. (1988): 'The Yorkshire and Lancashire lead mines. A study of lead mining in the South Craven and Rossendale Districts', *British Mining*, no. 33, pp. 1-68.

GILL, M. C. (1993): 'The Grassington Mines', *British Mining*, no. 46, *Monograph of the Northern Mines Research Society*.

GILMORE, P. (1882): *Alston Directory* (advert) (London: Porter).

GOLDRING, D. C. and D. A. GREENWOOD (1990): 'Fluorite mineralization at Beckermet iron ore mine, Cumbria, north England', *Transactions of the Institution of Mining and Metallurgy* (Section B: Applied Earth Science), vol. 99, pp. B113-B19.

GOODCHILD, J. G. (1875): 'Wulfenite at Caldbeck Fells', *Geological Magazine*, vol. 2, pp. 565-66.

GOODCHILD, J. G. (1881-82): 'Addenda to the list of minerals occurring in Cumberland and Westmorland: taken from the article by Bryce Wright in Jenkinson's larger guide to the Lake District', *Transactions of the Cumberland and Westmorland Association*, vol. 7, p. 178.

GOODCHILD, J. G. (1882): 'Contribution towards a list of the minerals occurring in Cumberland and Westmorland', *Transactions of the Cumberland Association for the Advancement of Literature and Science*, no. 7, pp. 101-203.

GOODCHILD, J. G. (1883): 'Contribution towards a list of the minerals occurring in Cumberland and Westmorland' (Part II), *Transactions of the Cumberland Association for the Advancement of Literature and Science*, no. 8, pp. 189-204.

GOODCHILD, J. G. (1885): 'Contribution towards a list of the minerals occurring in Cumberland and Westmorland' (concluding part), *Transactions of the Cumberland Association for the Advancement of Literature and Science*, no. 9, pp. 75-199.

GOODCHILD, J. G. (1889-90): 'Some observations upon the mode of occurrence and the genesis of metalliferous deposits', *Proceedings of the Geologists' Association*, vol. 11, pp. 45-69.

GOUGH, D. (1965): 'Structural analysis of ore shoots at Greenside lead mine, Cumberland, England', *Economic Geology*, vol. 60, pp. 1459-72.

GRAHAM, M. R. (2003): 'The reasons behind the increase and decrease in demand for fluorspar by the British steel industry', pp. 39-46 in R. A. FAIRBAIRN (ed.): *Fluorspar in the Pennines* (Killhope, Co. Durham: Friends of Killhope).

GRANT, D. (1985): 'The Sixth Duke of Somerset, Thomas Robinson and the Newlands Mines', *Transactions of the Cumberland and Westmorland Antiquaries and Archaeological Society*, vol. 85.

GRANTHAM, D. R. (1928): 'The petrology of the Shap granite', *Proceedings of the Geologist's Association*, vol. 39, pp. 299-331.

GRAY, G. and A. G. JUDD (2004): 'Barium sulphate production from mine waters in south east Northumberland', *British Mining*, no. 73, pp. 72-88.

GREEN, D. I. (1989): 'Scotlandite from Higher Roughton Gill, Caldbeck Fells, Cumbria', *Mineralogical Magazine*, vol. 53, p. 653.

GREEN, D. I. (1997): Supergene Cu, Pb, Zn and Ag minerals from Force Crag Mine, Coledale, Cumbria', *U.K. Journal of Rocks and Minerals*, vol. 18, pp. 10-14.

GREEN, D. I. (2003): 'Gearksutite from the Old Gang Mines, Swaledale, North Yorkshire, England', *Journal of the Russell Society*, vol. 8, pp. 41-43.

GREEN, D. I., T. F. BRIDGES and T. NEALL (2005): 'The occurrence and formation of galena in supergene environments', *Journal of the Russell Society*, vol. 8, pp. 66-70.

GREEN, D. I. and BRISCOE, P. J. (2002): 'Twenty Years in Minerals: The Classic Areas of Northern England', *The UK journal of Mines & Minerals*, no. 22, pp. 3-42.

GREEN, D. I., C. M. LEPPINGTON and T. NEALL (2003): 'Gartrellite, a mineral new to Britain from Low Pike, Caldbeck Fells, Cumbria, England', *Journal of the Russell Society*, vol. 8, pp. 43-44.

GREEN, D. I., T. NEALL, T. COTTERELL and C. M. LEPPINGTON (2003): 'Symplesite and parasymplesite from Cumbria and Cornwall, England', *Journal of the Russell Society*, vol. 8, pp. 16-17.

GREEN, D. I., T. NEALL and A. G. TINDLE (2005): 'The composition of fülöppite from Wet Swine Gill, Caldbeck Fells, Cumbria', *Journal of the Russell Society*, vol. 8, pp. 101-102.

GREEN, D. I. and J. R. NUDDS (1998): 'Harmotome from Brownley Hill mine, Nenthead, Cumbria', *Journal of the Russell Society*, vol. 7, pp. 31-33.

GREEN, D. I., M. S. RUMSEY, T. F. BRIDGES, A. G. TINDLE and R. A. IXER (2006): 'A review of the mineralisation at the Driggith and Sandbed mines, Caldbeck Fells, Cumbria', *Journal of the Russell Society*, vol. 9, pp. 4-38.

GREEN, D. I., A. G. TINDLE, T. F. BRIDGES and T. NEALL (2006): 'A review of the mineralization at Arm O'Grain, Caldbeck Fells, Cumbria', *Journal of the Russell Society*, vol. 9, pp. 44-53.

GREEN, D. I. and B. YOUNG (2006): 'Hydromagnesite and dypingite from the Northern Pennine Orefield, Northern England', *Proceedings of the Yorkshire Geological Society*, vol. 56, pp. 151-54.

GREEN, D. I., B. YOUNG and R. A. IXER (2006): 'Beaverite from Cumbria and Yorkshire', *Proceedings of the Yorkshire Geological Society*, vol. 56, pp. 155-58.

GREEN, J. F. N. (1915): 'The garnets and streaky rocks of the English Lake District', *Mineralogical Magazine*, vol. 17, 207-17.

GREENWELL, G. C. (1865-66): 'On the haematite mines of the Ulverston district', *Transactions of the Manchester Geological Society*, vol. 5, pp. 248-52.

GREENWOOD, D. A. and F. W. SMITH (1977): 'Fluorspar mining in the northern Pennines', *Transactions of the Institution of Mining and Metallurgy*, vol. 86, B181-90.

GREG, R. P. AND W. G. LETTSOM (1858): *Manual of the Mineralogy of Great Britain and Ireland* (London: John Van Voorst; reprinted 1977, Broadstairs, Kent: Lapidary Publications).

GUNN, W. (1900): 'The geology of Belford, Holy Island and the Farne Islands', *Memoir of the Geological Survey of Great Britain*.

GUTHRIE, R. G. (1966): 'Hurst Lead Mine, Swaledale', *Memoirs of the Northern Cavern and Mines Research Society*, vol. 1(5), pp. 44-45.

HACKER, D. (2003): 'Fluorite – the dealers and the collectors', pp. 16-23 in R. A. FAIRBAIRN (ed.): *Fluorspar in the Pennines* (Killhope, Co. Durham: Friends of Killhope).

HALL, T. M. (1868): *The mineralogists' directory* (London: Edward Stanford).

HALLIMOND, A. F. (1925): 'Iron ores: bedded ores of England and Wales. Petrography and Chemistry', Special Report on the Mineral Resources of Great Britain, *Memoir of the Geological Survey of Great Britain*, vol. 29.

HANCOX, E. G. (1934): 'Witherite and darytes', *Mining Magazine*, vol. 51, pp. 76-79.

HARE, T. H. (1844): *Series of Views of the Collieries in the Counties of Northumberland and Durham* (London: James Madden; reprinted 1969, Newcastle upon Tyne: Frank Atkinson).

HARKER, A. (1902): 'Notes on the Igneous Rocks of the English Lake District', *Proceedings of the Yorkshire Geological Society*, vol. 14, pp. 487-93.

HARKER, A. and J. E. MARR (1891): 'The Shap Granite and associated rocks', *Quarterly Journal of the Geological Society of London*, vol. 47, pp. 266-328.

HARKER, R. S. (1966): 'Copper Mines of the Lake District', *Memoirs of the Northern Cavern and Mines Research Society*, vol. 1(5), pp. 35-40.

HARLEY, S. (1999): *In the Bewick Vein, the story of a Northumberland lead mine* (Haydon Bridge, Northumberland: Honeycrook Press).

HARRIS, A. (1970): *Cumberland Iron: the story of Hodbarrow Mine 1855-1968* (Truro: D. Bradford Barton).

HARRIS, P. and G. W. DAGGER (1987): 'The intrusion of the Carrock Fell Gabbro Series (Cumbria) as a sub-horizontal tabular body', *Proceedings of the Yorkshire Geological Society*, 46, pp. 371-80.

HARRISON, R. K. (1975): 'Jasperoid and chalcedonic concretions in anhydrite, Sandwith Mine, St Bees, Cumbria', *Bulletin of the Geological Survey of Great Britain*, no. 52, pp. 55-60.

HARTLEY, E. G. J. (1900): 'On the constitution of the natural arsenates and phosphates – Part III: plumbogummite and hitchcockite', *Mineralogical Magazine*, vol. 12, 223-38.

HARTLEY, J. (1959/60): 'Coronadite from Cumberland', *Mineralogical Magazine*, vol. 32, pp. 343-44.

HARTLEY, J. (1984): 'A list of minerals associated with the ore deposits of the Caldbeck Fells, Cumbria', *Transactions of the Leeds Geological Association*, vol. 10, pp. 22-39.

HARWOOD, G. M. and F. W. SMITH (1986): 'Mineralization in Upper Permian carbonates at outcrop in Eastern England', in G. M. HARWOOD and D. B. SMITH (eds): *The English Zechstein and Related Topics*, pp. 103-111, Geological Society Special Publication.

HAWKES, L. and J. A. SMYTHE (1935): 'Ankerites of the North-umberland Coalfield', *Mineralogical Magazine*, vol. 24, pp. 65-75.

HESSENBERG, F. (1870): 'Caledonit von Red Gill, Cumberland', *Abhandlungen der Senckenbergische Naturforschende Gesellschaft, Frankfurt Am Main*, vol. 7, pp. 304-308.

HEWER, R. and A. McFADZEAN (1992): 'Iron', in *Cumbria Amenity Trust Mining History Society. Beneath the Lakeland Fells, Cumbria's Mining Heritage* (Cumbria: Red Earth Publications).

HEY, M. H. (1962): *An index of mineral species and varieties* (London: British Museum [Natural History]).

HEY, M. H. (1963): An appendix to the 2nd edition of *An index of mineral species and varieties* (London: British Museum [Natural History]).

HEY, M. H. (1974): A 2nd appendix to the 2nd edition of *An index of mineral species and varieties* (London: British Museum [Natural History]).

HEYES, W. F. (1997): 'The Formative Years of the Weardale Iron Company – Part 1: A Review of the Weardale Iron Mining Leases', *Bonny Moor Hen*, no. 9, pp. 21-31.

HIBBERT, A., W. H. JOHNSON and J. J. ADAMS (1940): *Survey of iron ore resources of West Cumberland* (Cumberland Development Council Ltd).

HILL, J. A. and K. C. DUNHAM (1968): 'The barytes deposit at Closehouse, Lunedale, Yorkshire', *Proceedings of the Yorkshire Geological Society*, vol. 36, pp. 351-72.

HIRST, D. M. and K. C. DUNHAM (1963): 'Chemistry and petrography of the Marl Slate of Durham, England', *Economic Geology*, vol. 58, pp. 912-40.

HIRST, D. M. and F. W. SMITH (1974): 'Controls of baryte mineralisation in the Lower Magnesian Limestone of the Ferryhill area, County Durham', *Transactions of the Institution of Mining and Metallurgy*, vol. 83, pp. B49-56.

HITCHEN, C. S. (1934): 'The Skiddaw granite and its residual products', *Quarterly Journal of the Geological Society of London*, vol. 90, pp. 158-200.

HOCHLEITNER, R. (1984): 'Kampylite Von Cumberland', *Lapis*, vol. 9, p. 27.

HODGE, B. L. (1970): 'The UK fluorspar industry and its basis', *Industrial Minerals*, April 1970, pp. 23-37.

HODGSON, E. (1870): 'On the situation of the iron-ore fossils in the Water Blain mines, South Cumberland, *Geological Magazine*, vol. 7, pp. 113-15.

HOLLAND, E. G. (1967): *Underground in Furness* (Clapham [via Lancaster]: Dalesman Books).

HOLLAND, E. G. (1981): *Coniston Copper Mines, a field guide* (Milnthorpe, Cumbria: Cicerone Press).

HOLLAND, E. G. (1986): *Coniston Copper* (Milnthorpe, Cumbria: Cicerone Press).

HOLLAND, J. G. and R. S. J. LAMBERT (1970): 'Weardale Granite', in G. A. L. JOHNSON (ed.): 'Geology of Durham County', *Transactions of the Natural History Society of Northumberland, Durham and Newcastle Upon Tyne*, vol. 41, pp. 103-118.

HORNSHAW, T. R. (1975): 'Copper mining in Middleton Tyas', *North Yorkshire County Record Office Publication*, no. 6.

HORNSHAW, T. R. (1976): 'The Richmond copper mine', *Journal of the North Yorkshire County Record Office*, no. 3, pp. 77-85.

HOUSTON, W. J. (1963): 'Rampgill Mine', *Mine and Quarry Engineering*, vol. 29, pp. 75-78.

HOWIE, R. A., E. PEGRAM and J. N. WALSH (1982): 'The content of rare earth elements in English Fluorites', *Journal of the Russell Society*, vol. 1, pp. 22-25.

HUDSON, R. G. S. (1938): 'The Harrogate mineral waters', *Proceedings of the Geologists' Association*, vol. 49, pp. 349-52.

HUNT, C. J. (1984): *The lead miners of the northern Pennines in the eighteenth and nineteenth centuries* (Newcastle upon Tyne: Davis).

HUNT, R. J. (1887): *British Mining* (London: Crosby Lockwood).

HUTCHINGS, W. M. (1898): 'The contact rocks of the Great Whin Sill', *Geological Magazine*, vol. 35, pp. 69-82 and 121-31.

HUNTER, R. H. and N. BOWDEN (1990): 'Itinerary 3: The Carrock Fell Igneous Complex', in F. MOSELEY (ed.): *The Lake District*, Geologists' Association Guide, pp. 57-67.

INESON, P. R. (1968): 'The petrology and geochemistry of altered quartz-dolerite in the Close House Mine area', *Proceedings of the Yorkshire Geological Society*, vol. 36, pp. 373-84.

INESON, P. R. (1969): 'Trace element aureoles in limestone wallrocks adjacent to lead-zinc-baryte-fluorite mineralization in the northern Pennine and Derbyshire ore fields', *Transactions of the Institution of Mining and Metallurgy*, vol. 78, pp. B29-40.

INESON, P. R. (1972): 'Alteration of the Whin Sill adjacent to baryte-witherite mineralization, Settlingstones Mine, Northumberland', *Transactions of the Institution of Mining and Metallurgy*, vol. 81, pp. B67-72.

INESON, P. R. (1976): 'Ores of the northern Pennines, the Lake District and North Wales', pp. 197-230 in K. H. WOLF (ed.): *Handbook of strata-bound and stratiform ore deposits*, vol. 5, *Regional studies* (Amsterdam: Elsevier).

INESON, P. R. and J. G. MITCHELL (1974): 'K-Ar Isotopic age determinations from some Lake District mineral localities', *Geological Magazine*, vol. 111, pp. 521-37.

INESON, P. R. and B. YOUNG (1994; reprinted 1996): 'Old Gang Mines, Hard Level, Gill, Friarfold Rake and Gunnerside Gill', in C. T. SCRUTTON (ed.): *Yorkshire Rocks and Landscape*, Yorkshire Geological Society (Cumbria: Ellenbank Press).

IVIMEY-COOK, H. C., G. WARRINGTON, N. E. WORLEY, S. HOLLOWAY and B. YOUNG (1995): 'Rocks of Late Triassic and Early Jurassic age in the Carlisle Basin, Cumbria (north-west England)', *Proceedings of the Yorkshire Geological Society*, vol. 50, pp. 305-316.

IXER, R. A. (1978): 'The distribution of bravoite and nickeliferous marcasite in central Britain', *Mineralogical Magazine*, vol. 42, pp. 149-50.

IXER, R. A. (1986): 'The ore mineralogy and paragenesis of the lead-zinc-fluorite-baryte orefields of the English Pennines and Mendip Hills', pp. 179-211 in J. R. CRAIG (ed.): *Mineral Parageneses* (Athens: Theophrastus).

IXER, R. A. and C. J. STANLEY (1987): 'A silver-nickel-cobalt mineral association from Tynebottom Mine, Garrigill, near Alston, Cumbria', *Proceedings of the Yorkshire Geological Society*, vol. 46, pp. 133-39.

IXER, R. A. and C. J. STANLEY (1998): 'Enargite group minerals from Scaleber Bridge, North Yorkshire, England', *Journal of the Russell Society*, vol. 7, pp. 41-42.

IXER, R. A., C. J. STANLEY and D. J. VAUGHAN (1979): 'Cobalt-, nickel- and iron sulpharsenides from the north of England', *Mineralogical Magazine*, vol. 43, pp. 389-95.

IXER, R. A., A. G. TINDLE, D. LISS and A. D. CHAMBERS (2005): 'On the occurrence of ilvaite in the pegmatitic facies of the Whin Sill, Teesdale, North Pennines, England', *U.K. Journal of Rocks and Minerals*, no. 25, pp. 39-46.

IXER, R. A., P. TURNER and B. WAUGH (1979): 'Authigenic iron and titanium oxides in Triassic red beds (St Bees Sandstone), Cumbria, Northern England', *Geological Journal*, vol. 14, pp. 179-92.

IXER, R. A. and D. J. VAUGHAN (1993): 'Lead-zinc-fluorite-baryte deposits of the Pennines, North Wales and the Mendips', pp. 355-418 in R. A. PATRICK and D. A. POLYA (eds): *Mineralisation in the British Isles* (London: Chapman and Hall).

IXER, R. A., B. YOUNG and C. J. STANLEY (1996): 'Bismuth-bearing assemblages from the Northern Pennine Orefield', *Mineralogical Magazine*, vol. 60, pp. 317-24.

JACKSON, D. I. and H. JOHNSON (1996): 'Lithostratigraphic nomenclature of the Triassic, Permian and Carboniferous of the UK offshore East Irish Sea Basin' (Nottingham: BGS).

JACKSON, H. (2001): *Shell Houses and Grottoes* (Shire Publications).

JACKSON, P. (1969): 'Some preliminary notes on the mining region of Alston Moor', *Memoirs of the Northern Cavern and Mines Research Society*, pp. 41-53.

JARS, G. (1765): *Voyages metallurgiques*, vol. 3, pp. 72-75 (Paris).

JENKINS, R. (1938): 'The society for the Mines Royal and the German colony in the Lake District', *Transactions of the Newcomen Society*, vol. 18, pp. 225-34.

JOHNSON, E. W., D. E. G. BRIGGS and J. L. WRIGHT (1996): 'Lake District pioneers – the earliest footprints on land', *Geology Today*, vol. 12, 147-51.

JOHNSON, E. W., N. J. SOPER, I. C. BURGESS, D. F. BALL, B. BEDDOE-STEPHENS, R. M. CARRUTHERS, N. J. FORTEY, S. HIRONS, J. W. MERRITT, D. MILLWARD, B. ROBERTS, A. B. WALKER and B. YOUNG (2001): 'Geology of the country around Ulverston', *Memoir of the British Geological Survey, England and Wales*, Sheet 48.

JOHNSON, G. A. L. (1961): 'Skiddaw Slates proved in the Teesdale Inlier', *Nature* (London), vol. 90, 996-97.

JOHNSON, G. A. L. and K. C. DUNHAM (1963): *The Geology of Moorhouse*, Monograph of the Nature Conservancy.

JOHNSON, G. A. L. and G. HICKLING (ed.) (1970): 'Geology of Durham County', *Transactions of the Natural History Society of Northumberland, Durham and Newcastle Upon Tyne*, vol. 41, pp. 1-158.

JOHNSON, G. A. L., J. R. NUDDS and D. ROBINSON (1980): 'Carboniferous stratigraphy and mineralisation at Ninebanks, West Allendale, Northumberland', *Proceedings of the Yorkshire Geological Society*, vol. 43, pp. 1-16.

JOHNSON, G. A. L. and B. YOUNG (1998): 'The trees which turned to stone', *Durham Town and Country*, issue 23, 41-43.

JOHNSTON, J. F. W. (1835): 'On the dimorphism of barytocalcite', *Philosophical Magazine*, vol. 6, pp. 1-4.

JOHNSTON, J. F. W. (1837): 'On the composition of the right rhombic barytocalcite the bicalcareo-carbonate of baryta of Dr Thomson', *Philosophical Magazine*, vol. 10, pp. 373-76.

JOHNSTON, J. F. W. (1839): 'On the composition of certain mineral substances of organic origin – Parts VI-VIII: Mineral resins', *Philosophical Magazine*, vol. 14, pp. 87-95.

JONES, K. and D. M. HIRST (1969): 'The distribution of barium, lead and zinc in the Lower and Middle Magnesian Limestones of Co. Durham', *Chemical Geology*, vol. 10, pp. 223-36.

JONES, N. S., P. D. GUION, D. DICKENS and B. YOUNG (1990): 'Coal Measures palaeoenvironments in West Cumbria. Report of Field Meeting', *Proceedings of the Yorkshire Geological Society*, vol. 48, pp. 119-20.

JONES, T. A. (1915): 'On the presence of tourmaline in Eskdale (Cumberland)', *Proceedings of the Liverpool Geological Society*, vol. 12, pp. 139.

KELLY, D. (1994): *The Red Hills. Iron mines of West Cumberland* (Ulverston, Cumbria: Red Earth Publications).

KENDALL, J. D. (1873-75): 'The haematite deposits of Whitehaven and Furness', *Transactions of the Manchester Geological Society*, vol. 13, pp. 231-83.

KENDALL, J. D. (1876): 'Haematite in the Silurians', *Quarterly Journal of the Geological Society of London*, vol. 32, pp. 180-83.

KENDALL, J. D. (1881-82): 'The haematite deposits of Furness', *Transactions of the North of England Institute of Mining and Mechanical Engineers*, vol. 31, pp. 211-37.

KENDALL, J. D. (1884): 'Mineral veins of the Lake District', *Transactions of the Manchester Geological Society*, vol. 17, pp. 292-341.

KENDALL, J. D. (1893): *The iron ores of Great Britain and Ireland* (London: Crosby Lockwood).

KENDALL, J. D. (1920): 'Lateral distribution of metallic minerals', *Mining Magazine*, vol. 23, pp. 75-80.

KENDALL, J. D. (1921): '"Flats" and "sops" in Furness', *Mining Magazine*, vol. 24, pp. 145-50.

KENDALL, P. F. and H. E. ROOT (1924): *Geology of Yorkshire* (Vienna: private publication).

KING, R. J. (1982): 'The Boltsburn Mine', *Mineralogical Record*, vol. 13, pp. 5-18.

KINGSBURY, A. W. G. (1957): 'New occurrences of phosgenite', *Mineralogical Magazine*, vol. 31, pp. 500-501.

KINGSBURY, A. W. G. (1958): 'Two minerals new to Britain: hubnerite and adamite', *Mineralogical Society*, notice no. 100.

KINGSBURY, A. W. G. and J. HARTLEY (1955): 'On the occurrence of the rare copper molybdate lindgrenite at Brandy Gill, Carrock Fell, Cumberland', *Mineralogical Magazine*, vol. 30, pp. 723-26.

KINGSBURY, A. W. G. and J. HARTLEY (1956): 'New occurrences of vanadium minerals (mottramite, descloizite and vanadinite) in the Caldbeck area of Cumberland', *Mineralogical Magazine*, vol. 31, pp. 289-95.

KINGSBURY, A. W. G. and J. HARTLEY (1956): 'Cosalite and other lead sulpho-salts at Grainsgill, Carrock Fell, Caldbeck, Cumberland', *Mineralogical Magazine*, vol. 31, pp. 296-300.

KINGSBURY, A. W. G. and J. HARTLEY (1956): 'Atacamite from Cumberland and Cornwall', *Mineralogical Magazine*, vol. 31, pp. 349-50.

KINGSBURY, A. W. G. and J. HARTLEY (1957): 'Childrenite from the Lake District, Cumberland', *Mineralogical Magazine*, vol. 31, p. 498.

KINGSBURY, A. W. G. and J. HARTLEY (1957): 'New occurrences of arseniosiderite', *Mineralogical Magazine*, vol. 31, pp. 499-500.

KINGSBURY, A. W. G. and J. HARTLEY (1957): 'New occurrences of rosasite in Britain', *Mineralogical Magazine*, vol. 31, pp. 501-502.

KINGSBURY, A. W. G. and J. HARTLEY (1957): 'Carpholite from Cumberland and Cornwall', *Mineralogical Magazine*, vol. 31, p. 502.

KINGSBURY, A. W. G. and J. HARTLEY (1957): 'Serpierite from the Lake District', *Mineralogical Magazine*, vol. 31, pp. 604-605.

KINGSBURY, A. W. G. and J. HARTLEY (1957): 'Beaverite from the Lake District', *Mineralogical Magazine*, vol. 31, pp. 700-702.

KINGSBURY, A. W. G. and J. HARTLEY (1958): 'Jarosite and natrojarosite from the Lake District', *Mineralogical Magazine*, vol. 31, pp. 813-15.

KINGSBURY, A. W. G. and J. HARTLEY (1960): 'Carminite and beudantite from the northern part of the Lake District and from Cornwall', *Mineralogical Magazine*, vol. 32, pp. 423-32.

KIRBY, G. A., H. A. BAILY, R. A. CHADWICK, D. J. EVANS, D. W. HOLLIDAY, S. HOLLOWAY, A. G. HULBERT, T. C. PHARAOH, N. J. P. SMITH, N. AITKENHEAD, B. BIRCH (2000): 'The structure and evolution of the Craven Basin and adjacent areas', *Subsurface Memoir of the British Geological Survey*.

KIRKSHAW, S. (1735): 'A letter concerning two pigs of lead', *Philosophical Transactions*, vol. 41, p. 560.

KNILL, D. C. (1960): 'Thaumasite from Co. Down, Northern Ireland', *Mineralogical Magazine*, vol. 32, pp. 416-18.

LAND, D. H. (1974): 'Geology of the Tynemouth district', *Memoir of the Geological Survey of Great Britain*.

LAWRENCE, D. J. D., C. L. VYE and B. YOUNG (2004): *A Geodiversity Audit For County Durham* (Durham County Council).

LAWRENCE, D. J. D., B. C. WEBB, B. YOUNG and D. E. WHITE (1986): 'The geology of the Late Ordovician and Silurian Rocks (Windermere Group) in the area around Kentmere and Crook', *Report of the British Geological Survey*, vol. 8, no. 5.

LE BAS, M. J. (1982b): 'The Caledonian granites and diorites of England and Wales', pp. 191-201 in D. S. SUTHERLAND (ed.): *Igneous rocks of the British Isles* (Chichester: John Wiley and Sons Ltd).

LEBOUR, G. A. (1886): *Outlines of the geology of Northumberland and Durham* (Newcastle upon Tyne: Lambert).

LEE, M. K. (1986): 'A new gravity survey of the Lake District and three-dimensional model of the granite batholith', *Journal of the Geological Society, London*, vol. 143, pp. 425-35.

LEITHART, J. (1838): *Practical observations on the mechanical structure, mode of formation, the repletion or filling up and the intersection and relative age of mineral veins* (London).

LEPPINGTON, C. M. and D. I. GREEN (1998): 'Antimonian claudetite from Wet Swine Gill, Caldbeck Fells, Cumbria, England', *Journal of the Russell Society*, vol. 7, pp. 36-37.

LEWIS, H. P. (1931): 'A sandstone with fluorspar cement and other sandstones from West Cumberland', *Geological Magazine*, vol. 68, pp. 543-57.

LIVINGSTONE, A. (1991): 'The zinc analogue of ktenasite from Smallcleugh and Brownley Hill mines, Nenthead, Cumbria', *Journal of the Russell Society*, vol. 4, pp. 13-15.

LIVINGSTONE, A. (2002): *Minerals of Scotland: Past and Present* (Edinburgh: NMS Publishing Limited/National Museums Scotland), pp. 212.

LIVINGSTONE, A., T. F. BRIDGES and R. E. BEVINS (1990): 'Schulenbergite and namuwite from Smallcleugh Mine, Nenthead, Cumbria', *Journal of the Russell Society*, vol. 3, pp. 23-24.

LIVINGSTONE, A. and P. E. CHAMPNESS (1993): 'Brianyoungite, a new mineral related to hydrozincite from the north of England orefield', *Mineralogical Magazine*, vol. 57, pp. 665-70.

LIVINGSTONE, A. and H. G. MACPHERSON (1983): 'Fifth Supplementary list of British Minerals (Scottish)', *Mineralogical Magazine*, vol. 47, pp. 99-105.

LLEWELLYN, P. G., S. A. MAHMOUD and R. STABBINS (1968): 'Nodular anhydrite in Carboniferous Limestone, West Cumberland', *Transactions of the Institution of Mining and Metallurgy* (Section B: Applied Earth Science), vol. 77, B18-B25.

LODGE, P. D. (1966): 'Hydraulic pumping and mining machinery, Sir Francis Level', *Memoirs of the Northern Cavern and Mine Research Society*, vol. 2, pp. 151-60.

LOUIS, H. (1917): 'Lead mines in Weardale, County Durham, worked by the Weardale Lead Company Limited', *Mining Magazine*, vol. 16, pp. 15-25.

LOUIS, H. and G. H. VELLACOTT, G. H. (1907): 'Mining: lead, iron, barytes and fluorspar', pp. 348-53 in W. PAGE (ed.): *Victoria History of County Durham*, vol. 2 (London: Constable).

LOVELAND, P. J. and V. C. BENDELOW (1984): 'Celadonite-aluminous-glauconite, an example from the Lake District, U.K.', *Mineralogical Magazine*, vol. 48, pp. 113-17.

LOWE, C. (1990): *Plants of Upper Teesdale* (Christopher Lowe).

LOWRY, D., A. J. BOYCE, R. A. D. PATTRICK, A. E. FALLICK and C. J. STANLEY (1991): 'A sulphur isotopic investigation of the potential sulphur sources for Lower Palaeozoic-hosted vein mineralization in the English Lake District', *Journal of the Geological Society, London*, vol. 148, pp. 993-1004.

LYON, R. J. P. and B. SCOTT (1957): 'Stratigraphical and structural ore controls on the Slitt Vein, at Heights Mine, Weardale, County Durham', *Transactions of the Institution of Mining and Metallurgy*, vol. 66, pp. 273-82.

McALLISTER, W. D. N. and R. J. FIRMAN (1980): 'Stream-sediment uranium anomalies in Eskdale, Cumbria, England', *Transactions of the Institution of Mining and Metallurgy*, vol. 89, pp. B177-81.

MACDONALD, W. (1925): 'The haematite deposits of West Cumberland', *Iron and Coal Trades Review*, vol. 110, p. 434.

McFADZEAN, A. (1989): *The Iron Moor* (Ulverston, Cumberland: Red Earth Publications).

MACPHERSON, H. (1983): 'References for and updating of L. J. Spencer's first and second supplementary lists of British Minerals', *Mineralogical Magazine*, vol. 47, pp. 243-57.

MACPHERSON, H. (1989): *Agates* (London: British Museum [Natural History]/Edinburgh: National Museums Scotland).

MACPHERSON, H. and A. LIVINGSTONE (1981): 'Glossary of Scottish mineral species', *Scottish Journal of Geology*, vol. 18, pp. 1-47.

MANNING, D. A. C., P. L. YOUNGER, F. W. SMITH, J. M. JONES, D. J. DUFTON and S. DISKIN (2007): 'A deep geothermal well at Eastgate, Weardale, UK: A novel exploration concept for low-enthalpy resources', *Journal of the Geological Society of London*, vol. 164, pp. 371-82.

MARR, J. E. (1900): 'Notes on the geology of the English Lake District', *Proceedings of the Geologists' Association*, vol. 16, pp. 449-83.

MARR, J. E. (1916): *The geology of the Lake District and the scenery as influenced by geological structure* (Cambridge: Cambridge University Press).

MATHESON, I. (1986): 'Coniston Mines in 1849', *The Mine Explorer*, vol. 2, pp. 41-44.

MELMORE, S. (1920): 'The metamorphism of the Carrock Fell Gabbro; with a note on the origin of the sulphide veins of the Caldbeck Fells', *Geological Magazine*, vol. 57, pp. 266-68.

MELMORE, S. (1943): 'Letters in the possession of the Yorkshire Philosophical Society', *North Western Naturalist*, vol. 18, p. 148.

MEYER, H. O. A. (1965): 'Revision of the Stratigraphy of the Permian Evaporites and Associated Strata in North-Western England', *Proceedings of the Yorkshire Geological Society*, vol. 35, pp. 71-89.

MIERS, H. A. (1889): 'Calcites from the neighbourhood of Egrement, Cumberland', *Mineralogical Magazine*, vol. 8, pp. 149-53.

MIERS, H. A. (1897): 'On some British pseudomorphs', *Mineralogical Magazine*, vol. 11, pp. 263-85.

MIERS, H. A. (1900): 'Note on hitchcockite, plumbogummite and beudantite analysed by Mr Hartley', *Mineralogical Magazine*, vol. 12, pp. 239-43.

MIERS, H. A. (1929): *Mineralogy* (2nd edition) (London: MacMillan).

MILBURN, T. A. (1987): *Life and Times in Weardale 1840-1910* (Weardale Museum: The Flying Printer, Norwich).

MILLER, N. (1982): *Heavenly Caves. Reflections on the Garden Grotto* (London: George Allen & Unwin).

MILLS, D. A. C. and D. W. HOLLIDAY (1998): 'Geology of the district around Newcastle Upon Tyne, Gateshead and Consett', *Memoir of the British Geological Survey*.

MILLS, D. A. C. and J. H. HULL (1976): 'Geology of the country around Barnard Castle', *Memoir of the Geological Survey of Great Britain*.

MILLWARD, D., B. BEDDOE-STEPHENS and B. YOUNG (1999): 'Pre-Acadian copper mineralisation in the English Lake District', *Geological Magazine*, vol. 136, pp. 159-76.

MILLWARD, D., E. W. JOHNSON, B. BEDDOE-STEPHENS, B. YOUNG, B. C. KNELLER, M. K. LEE and N. J. FORTEY (2000): 'Geology of the Ambleside District', *Memoir of the British Geological Survey*.

MILLWARD, D., F. MOSELEY and N. J. SOPER (1978): 'The Eycott and Borrowdale volcanic rocks', in F. MOSELEY (ed.): *The Geology of the Lake District*, Yorkshire Geological Society, Occasional Publication, vol. 3, pp. 99-120.

MILLWARD, D. and B. YOUNG (1990): 'Devoke Water and Yoadcastle', pp. 89-92 in F. MOSELEY (ed.): *Geology of the Lake District – Geologists' Association Guide* (London: Geologists' Association).

MILLWARD, D. and B. YOUNG (1990): 'Coniston to Lowes Water', pp. 119-24 in F. MOSELEY (ed.): *Geology of the Lake District – Geologists' Association Guide* (London: Geologists' Association).

MILNE, J. K., M. J. SAUNDERS and P. J. E. WOODS (1977): 'Iron-boracite from the English Zechstein', *Mineralogical Magazine*, vol. 41, pp. 404-406.

MITCHELL, J. G. and P. R. NESON (1975): 'Potassium-argon ages for the graphite deposits and related rocks of Seathwaite, Cumbria', *Proceedings of the Yorkshire Geological Society*, vol. 40, pp. 413-18.

MITCHELL, M., B. J. TAYLOR and W. H. C. RAMSBOTTOM (1978): 'Carboniferous', pp. 168-88 in F. MOSELEY (ed.): *The Geology of the Lake District*, Yorkshire Geological Society, Occasional Publication, no. 3.

MITCHELL, R. H. and H. R. KROUS (1971): 'Isotopic composition of sulphur and lead in galena from the Greenhow-Skyreholme area, Yorkshire, England', *Economic Geology*, vol. 66, pp. 243-51.

MITCHELL, W. R. (1977): 'Old Gang and Beyond', *Dalesman*, vol. 38, pp. 801-803.

MOON, J. (1964): 'Iron Ores of the Lake District and Furness', *Memoirs of the Northern Cavern and Mines Research Society*, vol. 1(3), pp. 11-14.

MOON, J. and J. D. J. WILDRIDGE. (1964): 'Embleton Lead Mine, Embleton, Cumbria', *Memoirs of the Northern Cavern and Mines Research Society*, vol. 1(2), pp. 14-15.

MOON, J. R. and J. D. J. WILDRIDGE (1968): 'The geology and non-ferrous mines of the Buttermere and Loweswater Valleys', *Memoirs of the Northern Cavern and Mine Research Society*, vol. 1(7), pp. 38-52.

MOORBATH, S. (1962): 'Lead isotope abundance studies on mineral occurrences in the British Isles and their Geological Significance', *Philosophical Transactions of the Royal Society*, vol. A254, pp. 295-359.

MORETON, S. (2005): 'Mud-filled fluorite vugs in Greenlaws Mine, Weardale, Co. Durham', *Journal of the Russell Society*, vol. 8, pp. 71-75.

MOSELEY, F. (1983): 'The Volcanic Rocks of the Lake District' (London: Macmillan Press).

MOSELEY, F. (1990): *Geology of the Lake District* (London: Geologists' Association).

MOSELEY, F. and D. MILLWARD (1982): 'Ordovician Volcanicity in the English Lake District', pp. 93-111 in D. S. SUTHERLAND (ed.): *Igneous rocks of the British Isles* (London: Wiley).

MUDDIMAN, E. W. (1940): *Witherite (natural barium carbonate) and its industrial uses*, a handbook issued jointly by Holmside and South Moor Collieries Ltd and the owners of Settlingstones Mines Ltd, Newcastle.

MUDDIMAN, E. W. (1942): 'Witherite', *Journal of the Oil and Colour Chemists Association*, vol. 25, pp. 127-42.

MURPHY, S. (1996): *Grey gold. Men, Mining and Metallurgy at the Greenside Lead Mine in Cumbria, 1825-1962* (Moiety Publishing).

NALL, W. (1888): *A Handbook to Alston* (Carlisle: Wordsworth Press).

NALL, W. (1904): 'The Alston mines', *Transactions of the Institute of Mining Engineers*, vol. 24, pp. 392-411.

NATURE CONSERVANCY COUNCIL (1987): 'Rescue collection of rare Lake District minerals', *Earth Science Conservation*, vol. 23, p. 36.

NEALL, T. and D. I. GREEN (2006): 'Scotlandite in the British Isles: its occurrence, morphology and paragenesis', *Journal of the Russell Society*, vol. 9, pp. 83-89.

NEALL, T. and C. M. LEPPINGTON (1994): 'Cinnabar from the Caldbeck Fells, Cumbria, England', *Journal of the Russell Society*, vol. 5, pp. 123-24.

NEALL, T. and C. M. LEPPINGTON (2003): 'New investigation of the mineralization in Burdell Gill, Caldbeck Fells, Cumbria', *Journal of the Russell Society*, vol. 8, pp. 1-9.

NEALL, T., A. G. TINDLE and D. I. GREEN (2006): 'The first British occurrence of arsendescloizite at Sandbed Mine, Caldbeck Fells, Cumbria', *U.K. Journal of Mines and Minerals*, no. 27, pp. 45-48.

NEALL, T., B. YOUNG, E. K. HYSLOP and R. D. FAKES (1993): 'Namibite from Buckbarrow Beck, Cumbria: the first British occurrence', *Mineralogical Magazine*, vol. 57, pp. 754-55.

NUTT, M. J. C. (1979): 'The Haweswater complex', pp. 727-33, in A. L. HARRIS, C. H. HOLLAND and B. E. LEAKE (eds): 'The Caledonides of the British Isles – reviewed', *Geological Society, London*, Special Publication, no. 8.

O'BRIEN, C., J. A. PLANT, P. R. SIMPSON and J. TARNEY (1985): 'The geochemistry, metasomatism and petrogenesis of the granites of the English Lake District', *Journal of the Geological Society*, vol. 142, pp. 1139-57.

OLIVER, R. L. (1956): 'The origin of garnets in the Borrowdale Volcanic Series and associated rocks, English Lake District', *Geological Magazine*, vol. 93, pp. 121-39.

ORCHARD, J. A. (1921): 'The mining, manufacture and uses of barites in the neighbourhood of Appleby, Westmorland', *Transactions of the Institution of Mining Engineers*, vol. 51, pp. 42-45.

OVE ARUP and PARTNERS (1990): 'North England', *Mining instability in Great Britain: Case Study Report*, Arup Geotechnics for the Department of the Environment, 1/ix (London: HMSO).

PARNELL, J. (1981): 'Genesis of the graphite deposit at Seathwaite, Borrowdale, Cumbria', *Geological Magazine*, vol. 119, pp. 511-12.

PATRICK, D. J. (1980): 'Mineral investigations at Carrock Fell, Cumbria – Part 1: Geophysical Survey', *Mineral Reconnaissance Programme Report, Institute of Geological Sciences*, no. 33.

PATTRICK, R. A. D. and D. A. POLYA (eds) (1993): *Mineralisation in the British Isles* (London: Chapman and Hall).

PEACH, B. N. and J. HORNE (1899): 'The Silurian rocks of Britain – Vol. 1: Scotland', *Memoir of the Geological Survey of the United Kingdom*.

PEEL, R. (1900): 'Notes upon the occurrence of barytes in a twenty-fathom fault at New Brancepeth Colliery', *The Colliery Manager J. Colliery Eng.*, vol. 16, pp. 56-58.

PENHALLURICK, R. D. (1997): 'The Mineral Collection of the Royal Institution of Cornwall', *U.K. Journal of Mines and Minerals*, no. 18, pp. 1-32.

PENNYPACKER, C. H. (1897): 'Calcite', *The Mineral Collector*, vol. 3, pp. 154-55.

PENNYPACKER, C. H. (1902): 'Barite', *The Mineral Collector*, vol. 9, pp. 116-18.

PENNYPACKER, C. H. (1904): 'A Trip to England', *The Mineral Collector*, vol. 2, pp. 65-66.

PHILLIPS, J. (1836): *Illustrations of the geology of Yorkshire; or a description of the strata and organic remains – Part 2: The Mountain Limestone District* (London: John Murray).

PHILLIPS, R. and F. W. SMITH (1974): 'Structural controls and palaeocirculation in fluorite veins in the north Pennine orefield, England', pp. 17-22 in *Symposium on Problems of Ore Deposition*, Varna, vol. 3.

PHILLIPS, W. (1819): *An elementary introduction to the knowledge of mineralogy* (London: William Phillips).

PHILLIPS, W. (1844): An elementary treatise on mineralogy (Boston: W. D. Ticknor and Co.).

PLANT, J., G. C. BROWN, P. R. SIMPSON and R. T. SMITH (1980): 'Signatures of metalliferous granites in the Caledonides', *Transactions of the Institution of Mining and Metallurgy*, vol. 89, pp. B198-210.

POOLE, G. (1937): 'The barytes, fluorspar and lead resources in Upper Teesdale and Weardale', *Report of the Technical Advisory Committee, North-East Development Board, Newcastle*.

POSTLETHWAITE, T. (1889-90): 'The deposits of metallic and other minerals surrounding the Skiddaw Granite', *Transactions of the Cumberland and Westmorland Association*, vol. 15, pp. 76-86.

POSTLETHWAITE, T. (1913): *Mines and mining in the English Lake District* (3rd edition) (Whitehaven: W. H. Moss; reprinted 1975 by Whitehaven: Michael Moon).

RAISTRICK, A. (1927): 'Lead mining and smelting in West Yorkshire', *Transactions of the Newcomen Society*, vol. 7, pp. 81-97.

RAISTRICK, A. (1930): 'A pig of lead with Roman inscriptions', *Yorkshire Archaeological Journal*, vol. 30, pp. 181-83.

RAISTRICK, A. R. (1934): 'The London Lead Company', *Transactions of the Newcomen Society*, vol. 14, pp. 119-62.

RAISTRICK, A. R. (1936): 'Lead smelting in the North Pennines during the seventeenth and eighteenth centuries', *Proceedings of the University of Durham Philosophical Society*, vol. 9, pp. 164-70.

RAISTRICK, A. (1936): '"Rare avis in terris"; The laws and customs of Lead Mines in West Yorkshire', *Proceedings of the University of Durham Philosophical Society*, vol. 9, pp. 180-90.

RAISTRICK, A. (1936): 'The copper deposits of Middleton Tyas, N. Yorkshire', *Naturalist*, pp. 111-15.

RAISTRICK, A. (1938): 'Mineral deposits in the Settle-Malham district, Yorkshire', *Naturalist*, pp. 119-25.

RAISTRICK, A. (1938): 'The Mineral Deposits', in R. G. S. HUDSON (ed.): 'Geology of the Country around Harrogate', *Proceedings of the Geologists' Association*, vol. 49, pp. 343-41.

RAISTRICK, A. R. (1938): *Two centuries of industrial welfare: The London (Quaker) Lead Company* (London: Society of Friends).

RAISTRICK, A. R. (1938): 'Vieille Montagne: a hundred years of smelting', *Transactions of the Newcomen Society*, vol. 18, pp. 266-67.

RAISTRICK, A. (1947-49): 'The Malham Moor mines, Yorkshire', *Transactions of the Newcomen Society*, vol. 26, pp. 69-77.

RAISTRICK, A. (1953): 'The lead mines of Upper Wharfedale', *Yorkshire Bulletin. Econ. Soc. Res.*, vol. 5, pp. 1-16.

RAISTRICK, A. (1954): 'The calamine mines, Malham, Yorks', *Proceedings of the University of Durham Philosophical Society*, vol. 2, pp. 125-30.

RAISTRICK, A. (1955): *Mines and miners of Swaledale* (Clapham [via Lancaster]: Dalesman Books).

RAISTRICK, A. (1955): 'The mechanisation of the Grassington Moor mines, Yorkshire', *Transactions of the Newcomen Society*, vol. 29, pp. 179-93.

RAISTRICK, A. (1966): 'Conistone Moor Mines, Wharfedale, Yorkshire', *Memoirs of the Northern Cavern and Mines Research Society*, vol. 1(5), pp. 27-32.

RAISTRICK, A. (1967): *Old Yorkshire dales* (Newton Abbot: David and Charles).

RAISTRICK, A. (1973): *Lead mining in the Mid Pennines. The mines of Nidderdale, Wharfedale, Airedale, Ribblesdale and Bowland* (Truro: D. Bradford Barton).

RAISTRICK, A. (1973): 'The London (Quaker) Company mines in Yorkshire', *Memoirs of the Northern Cavern and Mines Research Society*, vol. 2(3), pp. 127-32.

RAISTRICK, A. (1975): *The lead industry of Wensleydale and Swaledale* – Vol. 1: *The Mines*; Vol. 2: *The Smelting Mills* (Buxton: Moorland)

RAISTRICK, A. R. and B. JENNINGS (1965): *A history of lead mining in the Pennines* (London: Longmans).

RAISTRICK, A. and A. ROBERTS, A. (1984): *Life and work of the northern lead miner* (Stroud: Alan Sutton Publishing).

RANDALL, B. A. O. (1953): 'A large mineralised cavity in a tholeiite dyke in Northumberland', *Mineralogical Magazine*, vol. 30, pp. 246-53.

RANDALL, B. A. O. (1959): 'Stevensite from the Whin Sill in the region of the North Tyne', *Mineralogical Magazine*, vol. 32, pp. 218-25.

RANDALL, B. A. O. (1980): 'The Great Whin Sill and its associated dyke suite', pp. 67-75 in D. A. ROBSON (ed.): *The Geology of North East England*, Special Publication of the Natural History Society of Northumbria.

RANDALL, B. A. O. (1989): 'Dolerite-pegmatites from the Whin Sill near Barrasford, Northumberland', *Proceedings of the Yorkshire Geological Society*, vol. 47, pp. 249-65.

RANDALL, B. A. O. and J. M. JONES (1966): 'A recently discovered cavity in a dyke in the Northumberland coalfield', *Min. Engr.* vol. 125, pp. 309-17.

RANKIN, A. H. (1978): 'Macroscopic inclusions of fluid in British fluorites from the mineral collection of the British Museum (Natural History)', *Bulletin of the British Museum (Natural History), Geology Series*, vol. 30, pp. 297-307.

RANKIN, A. H. and M. J. GRAHAM (1988): 'Na, K and Li contents of mineralising fluids in the Northern Pennine Orefield, England and their genetic significance', *Transactions of the Institution of Mining and Metallurgy*, vol. 97, pp. B99-107.

RANSOMWE, T. (1848): 'Analysis of a saline spring in a lead mine near Keswick', *Memoir of the Manchester Literary and Philosophical Society*, vol. 8, pp. 399-401.

RASHLEIGH, P. (1797-1802): *Specimens of British minerals selected from the cabinet of Philip Rashleigh*, 2 vols (London: Bulmer).

RASTALL, R. H. (1942): 'The ore deposits of the Skiddaw district', *Proceedings of the Yorkshire Geological Society*, vol. 24, pp. 328-43.

RASTALL, R. H. and W. H. WILCOCKSON (1915): 'Accessory minerals of the Granitic rocks of the Lake District', *Quarterly Journal of the Geological Society, London*, vol. 71, pp. 592-622.

RIBBLESDALE, LORD (1807): 'Account of a mine of zinc ore

found in caverns at Malham Moors, and its application as paint', *Transactions of the Society of Arts*, vol. 35, pp. 35-38.

RIBBLESDALE, LORD (1807): 'Three letters giving an account of a mine of zinc ore and its application as paint', *Nat. Philos. Chem. Soc.*, vol. 21, pp. 12-14.

RICKARDS, R. B. (1978): 'Silurian', pp. 130-45 in F. MOSELEY (ed.): *The Geology of the Lake District*, Yorkshire Geological Society, Occasional Publication, no. 3.

RICHARDSON, D. T. (ed.) (1966): 'Spring Wood Level, Starbotton', *Northern Cavern and Mines Research Society*, Individual Survey Series, no. 1.

ROBERTS, D. E. (1983): 'Metasomatism and the formation of greisen in Gransgill, Cumbria, England', *Geological Journal*, vol. 18, pp. 43-52.

ROBERTSON, A. (2004): *The Walton Family* (Alston, Cumbria: Hundy Publications).

ROBSON, D. A. (1976): 'A guide to the geology of the Cheviot Hills', *Transactions of the Natural History Society of Northumbria*, vol. 43, pp. 1-23.

ROBSON, D. A. (ed.) (1980): *The geology of north east England* (Natural History Society of Northumbria).

ROGERS, C. (1994): *To be a gypsum miner* (Durham: Pentland Press).

ROGERS, P. J. (1974): 'Bravoite inclusions in fluorite from the Derbyshire and Askrigg ore-fields', *Bulletin of the Peak District Mines Historical Society*, vol. 5, p. 333.

ROGERS, P. J. (1978): 'Fluid inclusion studies on fluorite from the Askrigg Block', *Transactions of the Institution of Mining and Metallurgy*, vol. 87, pp. B125-32.

ROSE, W. C. C. and K. C. DUNHAM (1977): 'Geology and haematite deposits of south Cumbria', *Memoir of the British Geological Survey*.

ROWE, J., P. TURNER and S. BAILEY (1998): 'Palaeomagnetic dating of the west Cumbrian hematite deposits and implications for their mode of formation', *Proceedings of the Yorkshire Geological Society*, vol. 52, pp. 59-71.

RUDLER, F. W. (1905): *A handbook to a collection of the minerals of the British Islands mostly selected from the Ludlam Collection in the Museum of Practical Geology* (London: HMSO).

RUSKIN, J. (1879): *Deucalion, Collected studies of the lapse of waves and life of stones* (Orpington: George Allen).

RUSKIN, J. (1896): *Ethics of the Dust* (London: George Allen).

RUSSELL, A. (1918): *Report on the Occurrence of Witherite (BaCO₃) and Barytes (BaSO₄) at the Ushaw Moor Colliery, near Durham* (London: Library, Natural History Museum).

RUSSELL, A. (1925): 'A notice of the occurrence of native arsenic in Cornwall; of bismuthinite at Shap, Westmorland; and of smaltite and niccolite at Coniston, Lancashire', *Mineralogical Magazine*, vol. 20, pp. 299-304.

RUSSELL, A. (1927): 'Note of an occurrence of niccolite and ullmannite at the Settlingstones Mine, Fourstones, Northumberland', *Mineralogical Magazine*, vol. 21, pp. 383-87.

RUSSELL, A. (1936): 'Notes on the occurrence of Wulfenite at Brandy Gill, Cumberland, and of leadhillite at Drum-

ruck Mine, Kirkcudbrightshire', *Mineralogical Magazine*, vol. 24, pp. 321-23.

RUSSELL, A. (1944): 'Notes on some minerals either new or rare to Britain', *Mineralogical Magazine*, vol. 27, pp. 1-10.

RUST, S. A. (1991): 'Ktenasite from Smallcleugh Mine, Nenthead, Cumbria, the first British occurrence', *U.K. Journal of Mines and Minerals*, no. 9, p. 5.

RUST, S. A. (1997): 'Schulenbergite from Nentsberry Haggs mine, Cumbria', *U.K. Journal of Mines and Minerals*, no. 18, p. 15.

RYBACK, G. and P. C. TANDY (1992): 'Eighth supplementary list of British Minerals (English)', *Mineralogical Magazine*, vol. 56, pp. 261-75.

SAWKINS, F. J. (1966): 'Ore genesis in the north Pennine orefield in the light of fluid inclusion studies', *Economic Geology*, vol. 61, pp. 385-99.

SAWKINS, F. J., A. C. DUNHAM and D. M. HIRST (1964): 'Iron-deficient low temperature pyrrhotite', *Nature* (London), vol. 204, pp. 175-76.

SAY, P. J. and B. A. WHITTON (eds) (1981): 'Heavy metals in northern England: environmental and biological aspects', Proceedings of a Conference on Heavy Metals and the Environment, 17-18 October 1981, Department of Botany, University of Durham.

SCHNEIDER, H. W. (1884-85): 'On the haematite iron mines of Low Furness', *Transactions of the Cumberland Association*, Series 10, pp. 99-108.

SCHNELLMANN, G. A. (1947): 'The West Coast haematite bodies', *Mineralogical Magazine*, vol. 76, pp. 137-51.

SCRUTTON, C. (ed.) (1995): *Northumbrian Rocks and Landscape. A Field Guide* (Yorkshire Geological Society).

SCRUTTON, C. (ed.) (1996): *Yorkshire Rocks and Landscape. A Field Guide* (Yorkshire Geological Society).

SEAGER, A. F. and W. F. DAVIDSON (1952): 'Changes in habit during the growth of barite crystals from the north of England', *Mineralogical Magazine*, vol. 29, pp. 885-94.

SEDGWICK, A. (1846): 'The geology of the Lake District', in J. HUDSON (ed.): *Scenery of the Lakes* (3rd edition) (Kendal: J. Hudson).

SEDMAN, K. W. (1983): *The Mines and Minerals of Teesdale and Weardale* (Middlesbrough: Cleveland County Museum Service).

SELWYN-TURNER, J. (1963): 'The type locality for witherite', *Mineralogical Magazine*, vol. 33, pp. 431-32.

SHACKLETON, E. H. (1966): *Lakeland Geology* (Clapham [via Lancaster]: Dalesman Publishing Co.).

SHAW, J. L. (1880): 'On the haematite iron mines of the Furness District', *Proceedings of the Institution of Mechanical Engineers*, vol. 31, pp. 363-69.

SHAW, J. L. (1903): 'The probability of iron ore lying beneath the sands of the Duddon Estuary', *Journal of the Iron and Steel Institute*, Series 2, pp. 197-205.

SHAW, W. T. (1959): 'Joint discussion on the Lake District Mining field', pp. 219-23 in *The future of non-ferrous mining in Great Britain and Ireland* (London: Institution of Mining and Metallurgy).

SHAW, W. T. (1983): *Mining in the Lake Counties* (Clapham [via Lancaster]: Dalesman Books).

SHEPHERD, T. J. (1979): 'Microthermometric analysis of quartz-sulphide veins, Cumpston Hill, Yorkshire', *Isotope Geology Unit,* Report no. 79/3, Institute of Geological Sciences.

SHEPHERD, T. J., R. D. BECKINSALE, C. C. RUNDLE and J. DURHAM (1976): 'Genesis of Carrock Fell tungsten deposits Cumbria: fluid inclusion and isotopic study', *Transactions of the Institution of Mining and Metallurgy,* vol. 85, pp. B63-73.

SHEPHERD, T. J. and D. P. F. DARBYSHIRE (1981): 'Fluid inclusion Rb-Sr isochrons for dating mineral deposits', *Nature* (London), vol. 290, p. 5807.

SHEPHERD, T. J., D. P. F. DARBYSHIRE, G. R. MOORE and D. A. GREENWOOD (1982): 'Rare earth element and isoptopic geochemistry of the north Pennine ore deposits', *Bull. Bu. Rech. Geol. Min.* (2), Section II, no. 4, pp. 371-77.

SHEPHERD, T. J. and D. C. GOLDRING (1993): 'Cumbrian hematite deposits, north-west England, pp. 419-45 in R. A. D. PATRICK and D. A. POLYA (eds): *Mineralization in the British Isles* (London: Chapman and Hall).

SHEPHERD, T. J. and P. WATERS (1984): 'Fluid inclusion gas studies, Carrock Fell tungsten deposit, England: implications for regional exploration', *Mineralium Deposita,* vol. 19, pp. 304-14.

SHEPPARD, S. M. F. (1977): 'Identification of the origin of ore forming solutions by the use of stable isotopes', pp. 25-41 in *Volcanic processes in ore genesis* (London: Institution of Mining and Metallurgy and Geological Society).

SHERLOCK, R. L. (1919): 'Sundry unbedded iron ores of Durham, East Cumberland, North Wales, Derbyshire, the Isle of Man, Bristol District and Somerset, Devon and Cornwall', Special Reports on the Mineral Resources of Great Britain, *Memoir of the Geological Survey of Great Britain,* vol. 9.

SHERLOCK, R. L. (1921): 'Rock salt and brine', Special Reports on the Mineral Resources of Great Britain', *Memoir of the Geological Survey of Great Britain,* vol. 18.

SHERLOCK, R. L. and S. E. HOLLINGWORTH (1938): 'Gypsum and anhydrite, celestine and strontianite' (3rd edition), Special Reports on the Mineral Resources of Great Britain', *Memoir of the Geological Survey of Great Britain,* vol. 3.

SIDDIQUI, K. S. (1974): 'X-Ray analysis of specimens from the Hilton Borehole', Appendix B in I. C. BURGESS and D. W. HOLLIDAY (1974): 'The Permo-Triassic rocks of the Hilton Borehole, Westmorland – Part 1: Palaeogeography; Part 2: Petrology of the gypsum-anhydrite rocks', *Bulletin of the Geological Survey of Great Britain.*

SKIPSEY, E., B. C. WEBB and B. YOUNG (2002): *Rocks and Landscape of Alston Moor,* Cumbria Rigs and East Cumbria Countryside Project.

SMALL, A. T. (1978): 'Zonation of Pb-Zn-Cu-F-Ba mineralization in part of the North Yorkshire Pennines', *Transactions of the Institution of Mining and Metallurgy,* vol. 87, pp. 10-13.

SMALL, A. T. (1982): 'New data on tetrahedrite, tennantite,

chalcopyrite and pyromorphite from the Cumbrian and North Yorkshire Pennines', *Proceedings of the Yorkshire Geological Society,* vol. 44, pp. 153-58.

SMITH, A. (ed.) (2001): *The Rock Men. Pioneers of Lakeland Geology* (Cumberland Geological Society).

SMITH, B. (1924): 'Iron Ores: Haematites of West Cumberland, Lancashire and the Lake District', Special Reports on the Mineral Resources of Great Britain, *Memoir of the Geological Survey of Great Britain,* vol. 8.

SMITH, B. (1928): 'The origin of Cumberland Haematite', *Summary of Progress for the Geological Survey of Great Britain for 1927,* pp. 33-36.

SMITH, D. B. (1970): 'Permian and Trias', pp. 67-91 in 'Geology of Durham County', *Transactions of the Natural History Society of Northumberland, Durham and Newcastle Upon Tyne,* vol. 41.

SMITH, D. B. (1994): 'Geology of the country around Sunderland', *Memoir of the British Geological Survey.*

SMITH, D. B. and E. H. FRANCIS (1967): 'Geology of the country between Durham and West Hartlepool', *Memoir of the Geological Survey of Great Britain.*

SMITH, F. W. (1973): 'Supergene native copper in the northern Pennine orefield', *Mineralogical Magazine,* vol. 39, p. 244.

SMITH, F. W. (1974): 'Yttrium content of fluorite as a guide to vein intersections in partially developed fluorspar ore bodies', *Transactions of the Society of Mining Engineers,* vol. 255, pp. 95-96.

SMITH, F. W. (1977): 'The Weardale Lead Company Ltd', *British Mining.*

SMITH, F. W. (2003): 'The rise and fall of fluorspar mining in the Northern Pennines', pp. 24-38 in R. A. FAIRBAIRN (ed.): *Fluorspar in the Pennines* (Killhope, Co. Durham: Friends of Killhope).

SMITH, F. W. and R. G. HARDY (1981): 'Clay minerals in the veins of the North Pennine Orefield, U.K.', *Clay Minerals,* vol. 16, pp. 309-312.

SMITH, F. W. and R. PHILLIPS (1974): 'Temperature gradients and ore deposition in the north Pennine orefield', *Fortsch. Min.,* vol. 52, pp. 491-94.

SMITH, R. A. (1974): *A Bibliography of the Geology and Geomorphology of Cumbria* (Penrith: Cumberland Geological Society).

SMITH, R. A. (1990): *A Bibliography of the Geology and Geomorphology of Cumbria – Part 2 (1974-1990)* (Penrith: Cumberland Geological Society).

SMITH, S. (1923): 'Lead and zinc ores of Northumberland and Alston Moor', Special Reports on the Mineral Resources of Great Britain, *Memoir of the Geological Survey of Great Britain,* vol. 25.

SMITH, W. C. (1969): 'A History of the First Hundred Years of the Mineral Collection in the British Museum', *Bulletin of the British Museum (Natural History), Historical Series,* vol. 3, pp. 237-59.

SMITH, W. C. (1978): 'Early Mineralogy in Great Britain and Ireland', *Bulletin of the British Museum (Natural History), Historical Series,* vol. 6, pp. 49-74.

SMYTHE, J. A. (1921): 'Minerals of the north country: Fluorspar', *The Vasculum,* vol. 8, pp. 19-24.

SMYTHE, J. A. (1922): 'Minerals of the north country: Barium Minerals', *The Vasculum*, vol. 8, pp. 90-93 and 111-16.

SMYTHE, J. A. (1923): 'Minerals of the north country: Galena', *The Vasculum*, vol. 9, pp. 89-93 and 106-109.

SMYTHE, J. A. (1924): 'Minerals of the north country: Sulphides other than Galena', *The Vasculum*, vol. 11, pp. 7-14.

SMYTHE, J. A. (1924): 'Minerals of the north country: Silicates', *The Vasculum*, vol. 10. pp. 100-103.

SMYTHE, J. A. (1926): 'Minerals of the north country: (i) some general considerations; (ii) Calcium Carbonate', *The Vasculum*, vol. 12, pp. 140-47.

SMYTHE, J. A. (1927): 'Minerals of the north country: Oxides of iron and manganese', *The Vasculum*, vol. 13, 20-27.

SMYTHE, J. A. (1927): 'Minerals of the north country: Miscellaneous', *The Vasculum*, vol. 14, pp. 12-17.

SMYTHE, J. A. (1933): 'Nickel-Bearing goslarite, epsomite and melanterite from County Durham', *The Vasculum*, vol. 19, pp. 12-17.

SMYTHE, J. A. (1947): 'Chemical analysis of ullmannite from New Brancepeth, Co. Durham', *Mineralogical Magazine*, vol. 28, p. 173, pp. 65-75.

SMYTHE, J. A. and K. C. DUNHAM (1947): 'Ankerites and chalybites from the northern Pennine orefield and the north-east coalfield', *Mineralogical Magazine*, vol. 28, pp. 53-74.

SMYTHE, W. W. (1856): 'Iron ores of Great Britain – Part I', *Memoir of the Geological Survey of Great Britain*.

SOLLY, R. H. and A. HUTCHINSON (1889): 'Pseudomorphs of haematite after iron pyrites', *Mineralogical Magazine*, vol. 8, pp. 183-85.

SOLOMON, M. (1966): 'Origin of baryte in the north Pennine orefield', *Transactions of the Institution of Mining and Metallurgy*, vol. 75, pp. B230-31.

SOLOMON, M., T. A. RAFTER and K. C. DUNHAM (1971): 'Sulphur and oxygen isotope studies in the northern Pennines in relation to ore genesis', *Transactions of the Institution of Mining and Metallurgy*, vol. 80, pp. B259-76; Discussion, vol. 81, pp. B172-77, vol. 82, pp. 82, 46.

SOPER, N. J. and F. MOSELEY (1978): 'Structure', in F. MOSELEY (ed.): *The Geology of the Lake District*, Yorkshire Geological Society, Occasional Publication, no. 3, pp. 45-67.

SOPWITH, T. (1829): *Geological sections of Holyfield, Hudgill Cross Vein and Silver Band Lead Mines in Alston Moor and Teesdale* (Newcastle upon Tyne: E. Walker).

SOPWITH, T. (1832): 'On the application of isometric projection to geological plans and sections, with descriptive notices of the mining district at Nentsberry in the County of Cumberland', *Transactions of the Natural History Society of Northumberland, Durham and Newcastle Upon Tyne*, vol. 1, pp. 280-88.

SOPWITH, T. (1833): *An account of the mining district of Alston Moor, Weardale and Teesdale in Cumberland and Durham* (Alnwick: Davison).

SOWERBY, J. (1804-17): *British mineralogy: or coloured figures intended to elucidate the mineralogy of Great Britain* (London: published by the author).

SPENCER, L. J. (1898): 'Supplementary list of British minerals', *British Association for the Advancement of Science* (1899), pp. 875-77.

SPENCER, L. J. (1910): 'On the occurrence of alstonite and ullmannite in a barytes-witherite vein at New Brancepeth Colliery, near Durham', *Mineralogical Magazine*, vol. 15, pp. 302-11.

SPENCER, L. J. (1931): 'Second supplementary list of British minerals', *British Association for the Advancement of Science* (1932), p. 378.

SPENCER, L. J. (1948): 'Catalogue of topographical mineralogies and regional bibliographies', *Mineralogical Magazine*, vol. 28, pp. 301-32.

SPENCER, L. J. (1958): 'Third supplementary list of British minerals', *Mineralogical Magazine*, vol. 31, pp. 787-806.

STANLEY, C. J. and A. J. CRIDDLE (1979): 'Mineralization at Seathwaite Tarn, near Coniston, English Lake District: the first occurrence of wittichenite in Great Britain', *Mineralogical Magazine*, vol. 43, pp. 103-107.

STANLEY, C. J. and D. J. VAUGHAN (1980): 'Interpretative studies of copper mineralization to the south of Keswick, England', *Transactions of the Institution of Mining and Metallurgy*, vol. 89, pp. B25-30.

STANLEY, C. J. and D. J. VAUGHAN (1981): 'Native antimony and bournonite intergrowths in galena from the English Lake District', *Mineralogical Magazine*, vol. 44, pp. 257-60.

STANLEY, C. J. and D. J. VAUGHAN (1982): 'Copper, lead, zinc and cobalt mineralization in the English Lake District: classification, conditions of formation and genesis', *Journal of the Geological Society of London*, vol. 139, pp. 569-79.

STANLEY, C. J. and D. J. VAUGHAN (1982): 'Mineralization in the Bonser vein, Coniston, English Lake District: mineral assemblages, paragenesis, and formation conditions', *Mineralogical Magazine*, vol. 46, pp. 343-50.

STARKEY, R. E. (1987): 'Duftite from the North Pennine Orefield', *Proceedings of the Yorkshire Geological Society*, vol. 46, pp. 289.

STEPHENSON, D., R. E. BEVINS, D. MILLWARD, A. J. HIGHTON, I. PARSONS, P. STONE and W. J. WADSWORTH (1999): *Caledonian Igneous Rocks of Great Britain*, Geological Conservation Review, 17, Joint Nature Conservation Committee, Peterborough, 648 pp.

STRAHAN, A., J. S. FLETT and C. H. DINHAM (1917): 'Potash feldspar – phosphate of lime – alum shales – plumbago or graphite – molybdenite – chromite – talc and steatite (soapstone, soaprock and potstone) – diatomite', Special Reports on the Mineral Resources of Great Britain, *Memoir of the Geological Survey of Great Britain*, vol. 5.

STRENS, R. G. J. (1963): 'Some relationships between members of the epidote group', *Nature* (London), vol. 198, pp. 80-81.

STRENS, R. G. J. (1964): 'Pyromorphite as a possible primary phase', *Mineralogical Magazine*, vol. 33, pp. 722-23.

STRENS, R. G. J. (1964): 'Epidotes of the Borrowdale Volcanic rocks of central Borrowdale', *Mineralogical Magazine*, vol. 33, pp. 868-86.

STRENS, R. G. J. (1965): 'The graphite deposit of Seathwaite in Borrowdale, Cumberland', *Geological Magazine*, vol. 102, pp. 393-406.

SWEET, J. M. (1930): 'Notes on British Barytes', *Mineralogical Magazine*, vol. 22, pp. 257-70.

SWEET, J. M. (1960): 'British gold from the Bouglise collection', *Mineralogical Magazine*, vol. 32, pp. 420-21.

SYMES, R. F., M. M. WIRTH and B. YOUNG (1990): 'Schultenite from Caldbeck Fells, Cumbria: the first British occurrence', *Mineralogical Magazine*, vol. 54, p. 659.

TAYLOR, B. J., I. C. BURGESS, D. H. LAND, D. A. C. MILLS, D. B. SMITH and P. T. WARREN (1971): *British Regional Geology. Northern England* (4th edition) (London: HMSO, for Institute of Geological Sciences).

THOMPSON, L. M. (1933): 'The Great Sulphur Vein in Alston Moor' (K. C. DUNHAM [ed.]), *Proceedings of the University of Durham Philosophical Society*, vol. 9, pp. 91-98.

THOMSON, T. (1835): 'Account of some new species of minerals containing barytes', *Records of General Science*, vol. 1, pp. 369-75.

THOMSON, T. (1837): 'On the right-rhombic baryto-calcite', *Philosophical Magazine*, vol. 11, pp. 45-48.

THOMPSON, W. N. (1904): 'The Derwentwaters and the Radcliffes', *Transactions of the Cumberland and Westmorland Antiquarian and Archaeological Society* (new series), vol. 4, pp. 288-322.

TINDLE, A. G., T. F. BRIDGES and D. I. GREEN (2006): 'The composition of tsumebite from Roughton Gill Mine, Caldbeck Fells, Cumbria', *U.K. Journal of Mines and Minerals*, no. 27, pp. 48-50.

TRECHMANN, C. O. (1901): 'Note on a British occurrence of mirabilite', *Mineralogical Magazine*, vol. 13, pp. 73-74.

TRESTRAIL, G. F. W. (1931): 'The witherite deposit of the Settlingstones Mine, Northumberland', *Transactions of the Institution of Mining and Metallurgy*, vol. 40, pp. 56-65.

TRESTRAIL, G. F. W. (1938): 'Witherite in Northumberland', *Mine and Quarry Engineering*, vol. 3, pp. 247-51.

TROTTER, F. M. (1944): 'The age of the ore deposits of the Lake District and of the Alston Block', *Geological Magazine*, vol. 81, pp. 223-29.

TROTTER, F. M. (1945): 'The origin of the west Cumbrian haematites', *Geological Magazine*, vol. 82, pp. 67-80.

TROTTER, F. M. and S. E. HOLLINGWORTH (1932): 'The geology of the Brampton District', *Memoir of the Geological Survey of Great Britain*.

TROTTER, F. M., S. E. HOLLINGWORTH, T. EASTWOOD and W. C. C. ROSE (1937): 'Gosforth District', *Memoir of the Geological Survey of Great Britain*.

TURNBULL, L. (1975): *The history of lead mining in the North of England* (Newcastle upon Tyne; Harold Hill and Son).

TURNER, S. J. (1963): 'The Type Locality of Witherite', *Mineralogical Magazine*, vol. 33, pp. 431-32.

TYLECOTE, R. F. (1986): *The prehistory of metallurgy in the British Isles* (London: The Institute of Metals).

TYLER, I. (1990): *Force Crag, the history of a Lakeland mine* (Ulverston, Cumbria: Red Earth Publications).

TYLER, I. (1992): *Greenside, a tale of Lakeland miners* (Ulverston, Cumbria: Red Earth Publications).

TYLER, I. (1995): *Seathwaite wad and the mines of the Borrowdale valley* (Carlisle: Blue Rock Publications).

TYLER, I. (2000): *Gypsum in Cumbria* (Carlisle: Blue Rock Publications).

UGLOW, J. (2002): *The Lunar Men* (London: Faber & Faber).

VARVILL, W. W. (1920): 'Greenhow Hill lead mines', *Mining Magazine*, vol. 22, pp. 275-82.

VARVILL, W. W. (1937): 'A study of the shapes and distribution of the lead deposits in the Pennine limestones in relation to economic mining', *Transactions of the Institution of Mining and Metallurgy*, vol. 46, pp. 463-559.

VAUGHAN, D. J. and R. A. IXER (1980): 'Studies of sulphide mineralogy of north Pennine ores and their contribution to genetic models', *Transactions of the Institution of Mining and Metallurgy*, vol. 89, pp. B99-109.

VAUGHAN, D. J., M. SWEENEY, G. FRIEDRICH, R. DIEDEL and C. HARANCZYK (1983): 'The Kupferschiefer: An overview with an appraisal of the different types of mineralization', *Economic Geology*, vol. 84, pp. 1003-1027.

WADGE, A. J. (1978): 'Classification and stratigraphical relationships of the Lower Ordovician Rocks', in F. MOSELEY (ed.): *The Geology of the Lake District*, Yorkshire Geological Society, Occasional Publication, no. 3, pp. 68-78.

WADGE, A. J., J. D. APPLETON and A. D. EVANS (1977): 'Mineral investigations of Woodlands and Longlands in north Cumbria', *Mineral Reconnaissance Report of the Institute of Geological Sciences*, no. 14.

WADGE, A. J., J. H. BATESON and A. D. EVANS (1984): 'Mineral reconnaissance surveys in the Craven Basin', *Mineral Reconnaissance Report of the Institute of Geological Sciences*, no. 66.

WADGE, A. J., J. M. HUDSON, D. J. PATRICK, I. F. SMITH, A. D. EVANS, J. D. APPLETON and J. H. BATESON (1981): 'Copper mineralisation near Middleton Tyas', *Mineral Reconnaissance Report of the Institute of Geological Sciences*, no. 54.

WAGER, L. R. (1929): 'Metasomatism in the Whin Sill of the north of England – Part I: Metasomatism by lead vein solutions', *Geological Magazine*, vol. 66, pp. 97-110.

WAGER, L. R. (1929): 'Metasomatism in the Whin Sill of the north of England – Part II: Metasomatism by juvenile solutions', *Geological Magazine*, vol. 66, pp. 221-28.

WAINWRIGHT, A. (1987): *A Coast to Coast Walk from St Bees to Robin Hood Bay* (Middlesex: Penguin Books).

WALKER, E. E. (1904): 'Notes on the Garnet-bearing and associated Rocks of the Borrowdale Volcanic Series', *Quarterly Journal of the Geological Society of London*, vol. 60, pp. 70-104.

WALLACE, W. (1861): *The laws which regulate the deposition of lead ore in veins, illustrated by an examination of the geological structure of the mining districts of Alston Moor* (London: Stanford).

WALLACE, W. (1890): *Alston Moor, its pastoral people, its mines and miners* (Newcastle upon Tyne: Mawson, Swan and Morgan).

WALTON, J. (1946): 'The medieval mines of Alston', *Transactions of the Cumberland and Westmorland Antiquarian and Archaeological Society*, vol. 45, pp. 22-33.

WARD, J. (1756): 'Considerations of a draft of two large pieces of lead with Roman inscriptions upon them, found in Yorkshire', *Philosophical Transactions of the Royal Society*, vol. 49, pp. 686-700.

WARD, J. (1997): 'Early Dinantian evaporites of the Easton-1 Well Solway Basin, onshore, Cumbria, England', in N. S. MEADOWS, S. TRUEBLOOD, M. HARDMAN and G. COWAN (eds): *Petroleum Geology of the Irish Sea and Adjacent Areas*, Geological Society of London, Special Publication, 124.

WARD, J. C. (1876): 'The geology of the northern part of the English Lake District', *Memoir of the Geological Survey of Great Britain*.

WARD, J. C. (1877-88): 'Quartz as it occurs in the Lake District: its Structure and History', *Transactions of the Cumberland and Westmorland Association*, vol. 3, pp. 77-90.

WATSON, S. (1900): 'Recent mineral deposits and their relation to vein-formation', *Transactions of the Weardale Naturalists Field Club*, vol. 1, pp. 57-61.

WATSON, S. (1904): 'The Boltsburn Flats – their interest to the Student of Nature', *Transactions of the Weardale Naturalists Field Club*, vol. 1, pp. 146-50.

WATT, J. (1790): 'Some account of a mine in which the aerated barytes is found', *Memoirs of the Manchester Literary and Philosophical Society*, vol. 3, pp. 598-609.

WEIR, H. E. (1989): *Mining in Cumbria – A Bibliography* (Northern Mines Research Society).

WEIS, P. L., I. FRIEDMAN and J. P. GLEASON (1981): 'The origins of epigenetic graphite: evidence from isotopes', *Geochimica and Cosmochimica Acta*, vol. 45, pp. 2325-32.

WILDRIDGE, J. D. J. (1973): 'Mines of the Buttermere and Loweswater Valleys', *Memoirs of the Northern Cavern and Mines Research Society*, vol. 2(3), pp. 145-47.

WILKINSON, P. (2001): *The Nent Force Level and Brewery Shaft, Nenthead* (Cumbria North Pennines Heritage Trust).

WILLIAM OF MALMESBURY (c.690): *De Gestis Pontificum Anglorum* (London: Rolls Series, 1870).

WILLIAMSON, I. A. (1963): 'The Anglezarke Lead Mines', *Mining Magazine*, vol. 108, pp. 133-39.

WILLIAMSON, I. A. and P. M. WILLIAMSON (1968): 'Withering and witherite', *Pharmaceutical Journal* (10 August), p. 139.

WILSON, A. A. and J. D. CORNWELL (1982): 'The Institute of Geological Sciences borehole at Beckermonds Scar, North Yorkshire', *Proceedings of the Yorkshire Geological Society*, vol. 44, part 1, pp. 59-88.

WILSON, G. W, T. EASTWOOD, R. W. POCOCK, D. A. WRAY and T. ROBERTSON (1922): 'Barytes and witherite' (3rd edition), Special Reports on the Mineral Resources of Great Britain, *Memoir of the Geological Survey of Great Britain*, vol. 10.

WILSON, P. N. (1963): 'The Nent Force Level', *Transactions of the Cumberland and Westmorland Antiquarian and Archaeological Society*, vol. 63, pp. 253-80.

WILSON, W. E. (1994): 'The History of Mineral Collecting 1530-1799', *Mineralogical Record*, vol. 25, no. 6.

WIRTH, M. (1989): 'Native silver from Red Gill Mine, Caldbeck Fells, Cumbria', *Journal of the Russell Society*, vol. 2, p. 49.

WIRTH, M. (1995): 'Bavenite from Shap granite quarry, Cumbria, England', *Journal of the Russell Society*, vol. 6, p. 51.

WIRTH, M. M. and D. I. GREEN (2002): 'The Shap Granite Quarry, Cumbria', *U.K. Journal of Mines and Mining*, no. 22, pp. 43-52.

WITHERING, W. G. (1783): *Outline of mineralogy translated from the original of Sir Torbern Bergman* (Birmingham).

WITHERING, W. G. (1784): 'Experiments and Observations on the Terra Ponderosa' (communicated by Richard Kirwan), *Philosophical Transactions of the Royal Society*, vol. 74. pp. 293-311.

WOOD, M. (1991): 'Strontianite from Brownley Hill mine, Nenthead, Cumbria', *Journal of the Russell Society*, vol. 4, pp. 35-36.

WOOD, M. (1993): 'Strontianite from Gunnerside Gill, Swaledale, North Yorkshire', *U.K. Journal of Mines and Minerals*, no. 12, pp. 16-20.

WOODALL, F. D. (1975): *Steam Engines and Waterwheels* (Moorland Publishing).

WOODHALL, D. G. (2000): *Geology of the Keswick district*, British Geological Survey Sheet Description, England and Wales, sheet 29.

WOODWARD, J. (1695): *An essay towards a natural history of Earth*.

WOODWARD, J. (1728): 'A catalogue of the additional English native fossils in the collection of J. Woodward' (London: F. Fayram).

WOODWARD, J. (1729): *An attempt towards a natural history of the fossils of England: in a catalogue of the English fossils in the collection of J. Woodward M.D.* (London: F. Fayram).

WOOLER, E. (1926): 'Roman mining in Weardale', *Yorkshire Archaeological Journal*, vol. 28, p. 93.

WORDSWORTH, W. (1846): *The scenery of the Lakes including four letters on the geology of the Lake District by the Reverend Professor Sedgewick (edited by the Publisher)* (3rd edition) (Kendal: J. Hudson).

WRIGHT, B. M. (1879): 'List of minerals of the Lake District', in *Jenkinson's Guide to the Lakes* (6th edition).

WURTZBURGER, P. (1872): 'The haematite iron deposits of Furness', *Journal of the Iron and Steel Institute*, no. 1, pp. 135-42.

WYATT, J. (1995): *Wordsworth and the Geologists* (Cambridge University Press).

WYNNNE, I. N. (1943): 'The Stanhope Burn fluorite mine', *Mining Magazine*, vol. 69, pp. 265-73.

YOUNG, B. (1984): 'Siderite from the west Cumbrian hematite deposits', *Proceedings of the Cumberland Geological Society*, vol. 4, pp. 267-68.

YOUNG, B. (1984): 'Geology and history of Nab Gill Mine, Eskdale, Cumbria', *Proceedings of the Cumberland Geological Society*, vol. 4, pp. 269-75.

YOUNG, B. (1985): 'Mineralisation associated with the Eskdale intrusion, Cumbria', *Report of Programme Directorate A* (British Geological Survey).

YOUNG, B. (1985): 'Strontianite in the Northern Pennine Orefield', *Mineralogical Magazine*, vol. 49, p. 762.

YOUNG, B. (1985): 'Het erstgebied in de Northern Pennines en sijn mineralen' ['The Northern Pennine Orefield and its minerals'] (in Dutch), *Gea*, vol. 18, pp. 48-53.

YOUNG, B. (1985): 'The distribution of barytocalcite and alstonite in the Northern Pennine Orefield', *Proceedings of the Yorkshire Geological Society*, vol. 45, pp. 199-206.

YOUNG, B. (1985): 'The occurrence of pyromorphite in the Northern Pennines', *Journal of the Russell Society*, vol. 1, pp. 81-82.

YOUNG, B. (1986): 'Wulfenite from Ruthwaite Lodge, Grisedale – a new Lake District occurrence', *Proceedings of the Cumberland Geological Society*, vol. 4, pp. 403-404.

YOUNG, B. (1987): *Glossary of the minerals of the Lake District and adjoining areas* (Newcastle upon Tyne: British Geological Survey), 104 pp.

YOUNG, B. (1987): 'Uncommon Pennine minerals – Part 1: Aurichalcite in the Yorkshire and Cumbrian Pennines', *Transactions of the Leeds Geological Association*, vol. 11, pp. 25-32.

YOUNG, B. (1987): 'Uncommon Pennine minerals – Part 2: Strontianite from the Yorkshire and Pennines', *Transactions of the Leeds Geological Association*, vol. 2, pp. 33-40.

YOUNG, B. (1988): 'Wavellite and variscite from Scar Crag Cobalt Mine, Cumbria', *Proceedings of the Cumberland Geological Society*, vol. 5, pp. 13-16.

YOUNG, B. (1990): 'Nab Gill Mine, Eskdale', pp. 124-27 in F. MOSELEY (ed.): *Geology of the Lake District – Geologists' Association Guide* (London: Geologists' Association).

YOUNG, B. (1991): 'An occurrence of witherite in Weardale, Co. Durham, Northern Pennines', *Transactions of the Natural History Society of Northumbria*, vol. 55, pp. 206-207.

YOUNG, B. (1992): 'The Eskdale Granite', pp. 46-53 in M. DODD (ed.): *Lakeland Rocks and Landscape*, Cumberland Geological Society (Cumbria: Ellenbank Press).

YOUNG, B. (1993): 'Some new occurrences of barytocalcite in the Northern Pennine Orefield', *Transactions of the Natural History Society of Northumbria*, vol. 56, pp. 61-64.

YOUNG, B. (1994): 'Lead mining at Mosser: a geological note', pp. 483-84 in A. J. L. WINCHESTER (ed.): *The Diary of Isaac Fletcher of Underwood, Cumberland, 1756-1781*, Cumberland and Westmorland Antiquarian & Archaeological Society Extra Series, vol XXVII (Kendal).

YOUNG, B. (1995): 'The Northern Pennine Orefield: Weardale and Nenthead', pp. 154-161 in C. T. SCRUTTON (ed.): *Northumbrian Rocks and Landscape: A field guide* (Cumbria: Ellenbank Press and Yorkshire Geological Society).

YOUNG, B. (1995): 'Past wealth – present problems: Cumbrian Iron Ore', *Earthwise*, issue 6, March 1995, pp. 20-21.

YOUNG, B. (1998): 'Mines and minerals. The Scordale Lead Mines, Warcop', *Sanctuary*, no. 27, pp. 10-11.

YOUNG, B. (2001): 'Copper mineralisation in the Permian Magnesian Limestone at Raisby, Coxhoe, County Durham, England', *Journal of the Russell Society*, vol. 7, pp. 75-78.

YOUNG, B. (2003): 'Fluorspar in the ground', pp. 4-15 in R. A. FAIRBAIRN (ed.): *Fluorspar in the North Pennines* (Killhope, Co. Durham: Friends of Killhope, 2003).

YOUNG, B. (2005): 'A large glacial erratic boulder of gypsum from the Durham coast', *Transactions of the Natural History of Northumbria*, vol. 54, pp. 207-209.

YOUNG, B., S. M. ANSARI and R. J. FIRMAN (1988): 'Field relationships, mineralogy and chemistry of the greisens and related rocks associated with the Eskdale Granite, Cumbria', *Proceedings of the Yorkshire Geological Society*, vol. 47, pp. 109-123.

YOUNG, B. and M. P. BOLAND (1992): 'Geology and land-use planning: Great Broughton-Lamplugh area, Cumbria – Part 1: Geology', *British Geological Survey Technical Report*, no. WA/92/54.

YOUNG, B., M. P. BOLAND and E. K. HYSLOP (1992): 'Metavoltine from Cockermouth, Cumbria', *Mineralogical Magazine*, vol. 56, pp. 31-432.

YOUNG, B. and T. F. BRIDGES (1984): 'Harmotome from Northumberland', *Transactions of the Natural History Society of Northumbria*, vol. 52, pp. 24-26.

YOUNG, B., T. F. BRIDGES and E. K. HYSLOP (1994): 'Leadhillite from the Northern Pennine Orefield, England', *Journal of the Russell Society*, vol. 5, pp. 121-23.

YOUNG, B., T. F. BRIDGES and P. R. INESON (1987): 'Supergene cadmium mineralisation in the Northern Pennine Orefield', *Proceedings of the Yorkshire Geological Society*, vol. 46, pp. 275-78.

YOUNG, B. and A. H. COOPER (1988): 'The geology and mineralisation of Force Crag Mine, Cumbria', *Proceedings of the Cumberland Geological Society*, vol. 5, pp. 5-12.

YOUNG, B., M. P. COOPER, P. H. A. NANCARROW and D. I. GREEN (1990): 'Supergene minerals from Low Pike, Caldbeck Fells, Cumbria', *Journal of the Russell Society*, vol. 3, pp. 25-28.

YOUNG, B. and M. B. DODD (1984): 'The west Cumberland hematite orefield: 15th-17th July 1983 – Report of Field Meeting', *Proceedings of the Yorkshire Geological Society*, vol. 4, pp. 137-38.

YOUNG, B., A. DYER, N. HUBBARD and R. E. STARKEY (1991): 'Apophyllite and other zeolite-type minerals from the Whin Sill of the Northern Pennines', *Mineralogical Magazine*, vol. 55, pp. 203-207.

YOUNG, B., R. FAIRBAIRN and E. K. HYSLOP (1996): 'Coronadite from Weardale, Northern Pennines', *Journal of the Russell Society*, vol. 6, p. 97.

YOUNG, B., R. J. FIRMAN and R. E. STARKEY (1988): 'An occurrence of apophyllite at Shap, Cumbria', *Mineralogical Magazine*, vol. 52, pp. 415-16.

YOUNG, B. and I. FORBES (1997): 'A new lead mine in the northern Pennines', *Geology Today* (July-August 1997), pp. 154-57.

YOUNG, B., N. J. FORTEY and P. H. A. NANCARROW (1986): 'An

occurrence of tungsten mineralisation in the Eskdale Intrusion, West Cumbria', *Proceedings of the Yorkshire Geological Society*, vol. 46, pp. 15-21.

YOUNG, B., N. J. FORTEY and P. H. A. NANCARROW (1991): 'Russellite from Buckbarrow Beck, Cumbria, England', *Journal of the Russell Society*, vol. 4, pp. 3-7.

YOUNG, B. and J. G. FRANCIS (1989): 'Rosasite from West Pasture Mine, Weardale, Co. Durham', *Journal of the Russell Society*, vol. 2, p. 50.

YOUNG, B., L. GREENBANK and P. H. A. NANCARROW (1990): 'Alstonite in situ at Brownley Hill Mine, Nenthead, Cumbria', *Mineralogical Magazine*, vol. 54, pp. 515-16.

YOUNG, B., E. K. HYSLOP, T. F. BRIDGES and J. COOPER (2005): 'New records of supergene minerals from the Northern Pennine Orefield', *Transactions of the Natural History of Northumbria*, vol. 64, pp. 211-14.

YOUNG, B., E. K. HYSLOP, T. F. BRIDGES and N. HUBBARD (1998): 'Thaumasite from High Sedling Mine, Weardale, County Durham, and from Shap, Cumbria, England', *Journal of the Russell Society*, vol. 7, p. 40.

YOUNG, B., E. HYSLOP and D. MILLWARD (1992): 'Barium pharmacosiderite from Caldbeck Fells, Cumbria', *Journal of the Russell Society*, vol. 4, pp. 64-66.

YOUNG, B., P. R. INESON, T. F. BRIDGES and M. E. SMITH (1989): 'Cinnabar from the Northern Pennines, England', *Mineralogical Magazine*, vol. 53, pp. 388-90.

YOUNG, B. and E. W. JOHNSON (1985): 'Langite and posnjakite from the Lake District', *Journal of the Russell Society*, vol. 1, p. 80.

YOUNG, B. and D. J. D. LAWRENCE (2002): 'Geology of the Morpeth District – a brief explanation of the geological map,' *Sheet Explanation of the British Geological Survey*, 1:50 000, Sheet 14 Morpeth (England and Wales).

YOUNG, B., D. J. D. LAWRENCE, C. L. VYE and C. WOODLEY-STEWART (2004): *North Pennines Area of Outstanding Natural Beauty: A Geodiversity Audit and Action Plan 2004-09*, North Pennines Area of Outstanding Natural Beauty (AONB) Partnership.

YOUNG, B., A. LIVINGSTONE and N. THOMSON (1992): 'Fraipontite from Wensleydale, North Yorkshire', *Proceedings of the Yorkshire Geological Society*, vol. 49, pp. 25-127.

YOUNG, B., D. MILLWARD and D. C. COOPER (1992): 'Barium and base-metal mineralization associated with the southern margin of the Solway-Northumberland Trough. Conference Report', *Transactions of the Institution of Mining and Metallurgy*, vol. 101, pp. B171-72.

YOUNG, B. and P. H. A. NANCARROW (1988): 'Millerite from the Cumbrian Coalfield', *Journal of the Russell Society*, vol. 2, pp. 5-7.

YOUNG, B. and P. H. A. NANCARROW (1988): 'Rozenite and other sulphate minerals from the Cumbrian Coalfield', *Mineralogical Magazine*, vol. 52, pp. 551-52.

YOUNG, B. and P. H. A. NANCARROW (1990): 'Hausmannite from west Cumbria', *Journal of the Russell Society*, vol. 3, pp. 29-30.

YOUNG, B. and W. NICHOL (1997): 'A very rare mineral indeed', *The Northumbrian*, issue no. 40, pp. 40-43.

YOUNG, B., T. PETTIGREW and T. F. BRIDGES (1985): 'Rosasite and aurichalcite from the Northern Pennine Orefield', *Transactions of the Natural History Society of Northumbria*, vol. 54, pp. 31.

YOUNG, B., E. R. PHILLIPS and B. SMITH (2005): 'Fluorite-bearing marble from Barrasford Quarry, Northumberland', *Transactions of the Natural History of Northumbria*, vol. 64, pp. 199-205.

YOUNG, B. and A. PRINGLE (1999): 'Souvenirs of a Fiery Past', *The Northumbrian*, no. 48, February/March 1999, pp. 43-45.

YOUNG, B. and A. T. RAINE (1985): 'Thenardite (Na_2SO_4), a mineral new to Britain, from Sussex and Cumbria', *Journal of the Russell Society*, vol. 1, pp. 92-93.

YOUNG, B., G. RYBACK, R. S. W. BRAITHWAITE and J. G. FRANCIS (1997): 'Prosopite, doyleite and otavite from Coldstones Quarry, Pateley Bridge, North Yorkshire', *Mineralogical Magazine*, vol. 61, pp. 895-97.

YOUNG, B. and D. SCHOFIELD (1990): 'Stevensite from Upper Teesdale, Co. Durham, a Second British occurrence', *Transactions of the Natural History Society of Northumbria*, vol. 150, p. 150.

YOUNG, B., M. T. STYLES and N. G. BERRIDGE (1985): 'Niccolite-magnetite mineralization from Upper Teesdale, North Pennines', *Mineralogical Magazine*, vol. 49, pp. 555-59.

YOUNGER, P. L. (2003): 'Frazer's Grove: the life and death of the last great Weardale fluorspar mine', pp. 47-72 in R. A. FAIRBAIRN (ed.): *Fluorspar in the Pennines* (Killhope, Co. Durham: Friends of Killhope).

Index

The use of italics denotes an illustration, map or photographic reference.

Hadrian's Wall 2, 14, 43, 69
Haile Moor mine 82, *83*, 87, 118, 122
halite 15, 16
Handsome Mea Cross veins 54
Hard Level 76; Gill 75, 76, 77
Hard Rigg Edge 48
Hardberry Hill 67
Harehope Gill mine 44-45
Harkerside 72-73, 78
harmotome 28, 70
Harnisha Burn mine 44-45
Hartlepool 15
Hartley Birkett 72, 99; mine *72-73*
Hartley Castle lead mine 99
Hartside mine 44-45; Pass 48
Hartsop 22; Hall mine 22, 24, 37, 42
Harwood Beck 65; Valley 65, 66
Haug Langauer and Company of Augsburg 36
hausmannite 32, 85, 86, *162*
Haweswater 21; complex 9
Hawkshead 108
Haydon Bridge 69
Healeyfield mine 44-45
Heartycleugh mine *51*, 56
Hedworth Barium Company 66
Heights mine 27, 44-45, 62, 121, 153; Quarry 44-45, 62
Helder mine *83*, 86
Helvellyn 7, 21, 22, 41; mine *37*
hematite 32, 33, *84*, 85, 86, 87, 88, 91, 96, 118, 119, 122, 129, 130, 132, 136, 138, 140, 145, 162, 163, *167*, 168, 169; 'kidney ore' 32, 82, *84*, 85, 86, 88, 90, 91, 122, *163*, *164*; 'sop' orebody *33*; specular 32, 86; veins 82
hemimorphite 31, 41, 78, 79, 80, 81
Hesk Fell mine *37*
Hesleywell Hush *51*
Hexham 52, 69, 70,

102, 119, 123, 124, 174
High Crossgates mine *89*
High Crossgill mine *83*
High Cross vein 52, 124
High Cup Nick 14
Highfield Hushes 65
High Force 105; Quarry *45*, 67; waterfall *14*, 67
High House mine *83*
High Hundybridge mine *51*
High Level Gill (*see* Friarfold Hush)
High Longrigg 32, 72; mine *72-73*
High Mark 72, 80; mines *72-73*
Hilton 49, 121; Beck 49; mine 28, 45, 49, 50, 88
Hochstetter 43; methods 72
Hodbarrow mine *89*, 122
Hogget Gill 21, *37*, 40, 42
Holebeck mine *83*
Holyfield mine 30, *51*, 55
Holystone 11
Honister Quarry 8
Hope Level mine 44-45
hornblende 10
hornfels 20
hornfelsed mudstones 10
Hudeshope Beck 65, 68
Hudgill Burn: Cross vein 54; mine 30, *51*, 55
Hungry Hushes 72-73, 77
Hunstanworth 69
Hurst mines *72-73*, 77
Hurst Moor 77
hushes, hushing 53, 68, 75
hydroxyapophyllite 60
hydrozincite 30, 55

Iapetus Ocean and closure 5, 8
ICI (Imperial Chemical Industries) 48
ignimbrite 8
ikaite 16
ilmenite 9
Imperial Smelting Corporation 53
Ingleborough 13
Ingleton, Ingleton Group 5, 14

Ireshopeburn 149; mine 44-45, 60, 121
iron 2, 43, 126; ore 24, 25, 36, 48, 118

Jackass Level 72, *74*, 81 (*see* Greenhow Hill)
Jacktrees mine *83*
jamesonite 21, 42
Jamie vein 81
jarosite 31
jarrowite 16-17
Jarrow Slake 16
Jeffrey's mine 44-45
Jew Limestone 49, 63, 155
Jug vein 52, 124

Kathleen Pit 90
Keirsleywell Row 56
Keld Heads: mine 72-73, 79; Smelt Mill 79
Keldside mine 72-73
Kelton mine *37*, *83*; Quarry 32, 84
Kendal 108
Keswick 10, 22, 38, 39, 42, 97, 104, 105; grotto at *98*; pencil industry 20, 38
Kettlewell 79; Beck 80; Smelt Mill *72-73*
'kidney ore' (*see* hematite)
Kilhope Burn 59
Kilhope mines 59; Park Level 44, 59
Kinniside mine *37*
Kirkby Stephen 31, 32, 74
Kirkby Thore 15, 16
Kirkstone Pass 42; Quarry 8
Knock 49
Knockmurton 82; mines 32, *37*, 82, *83*, 122
Kupferschiefer 15

labradorite 7
Lady's Rake mine 28, 44-45, 65, 66
Lake District 1, 2, 5-6, 7, 8, 9, 10, 11, 12, 15, 16, 20, 21, 22, 23, 24, 32, 33, 36, *37* (map), 38-43, 72, 82, 87, 97, 101, 104, 105, 106, 120, 122, 148 and *passim*; Batholith 19, 23, 33; copper veins 23; galena 22; granites 9; lead mines 36; minerals of 96-97; veins 42

'Lakeland green slates' 8
Lamplugh 82
Lancashire 36, 123
Lane End mine *72-73*, 75
Langdale Pikes 7
Langdon: Beck 66; mines *45*
Langhorn Pit 85
Langley Barony mine 44-45, 70, 71
Langley: Castle 71; Smelt Mill 44-45, 71
Langthwaite 75, 76, 77; Octagon Smelt Mill *72-73*, 77
lapilli, accretionary 7
Laporte Company 49
laumontite 10
Laurentia, Laurentian continent 5; plates 8
Lead and Zinc Company 48
lead 2, 43, 47, 105, 119, 123; mining 24, 48, 94; ore 2, 25, 36, 38, 99; smelting 46
leadhillite 23, 30, 69
Lead-zinc veins 22-23
Leehouse Well mine *51*
Levelgate mine 44-45
Levers Water 43
Liassic rocks 16
Liddesdale 2
limestone 2, 11, 12, 13, 16, 29; concretionary ('cannon-ball limestone') 15; quarries 126, 136, 136, 149, 152, 155, 172
limonite, 'limonitic' 48, 60, 61, 118, 126, 127, 141, 148
linarite 23
Lindal Cote mine *89*
Lindal Moor 90; fault system 90; mine *89*, 90; vein 90
Linewath Bridge *37*
Lingmell Gill 8, *37*
Lintzgarth Smelt Mill 64
lithophone, stone (*see* xylophone, stone) 10
Little Gill mine *51*
Little Moor Foot 72; mine *72-73*
Little Town 39
Lodgesike mine *45*, 68; Lodgesike-Manorgill vein 67-68
Lolly mine *72-73*
London Lead Company ('Governor and Company for Smelting

Down Lead with Pit Coal and Sea Coal') 44, 46, 47, 48, 49, 52-53, 54, 55, 62, 65, 68, 69, 74, 78
Lonely Hearts orebody (*see* Florence mine)
Long Fell mine *45*
Long Work 39
Longcleugh mine 44-45, 56
Longlands Lake 83, 85; mine *83*, 85
Lonsdale mine *83*, 84
Loppysike mine 44
'loughs' (*see* 'flats' and 'vugs') 29, 32, 84, 85, 86, 97, 125
Lover Gill mine 72, 78
Lowfield mine *89*
Low Force waterfall 14
Low Hartsop mine *37*
Low Row 77
Lower Felltop Limestone 65
Lower Palaeozoic rocks 26, 30
Lownathwaite mine *72-73*
Lunedale 68; fault-system 68
Lunehead mine *45*, 68

Magnesian Limestone 3, 15, 29
magnetite 9, 10, 14, 21, 42, 43, 62, 65; skarn 28
Main Limestone 31
malachite 22, 30, 31, 32, 39, 75, 85, 91
Malham 8, 31, 32, 72-73, 74; manganese 65, 162; ore 36; oxide minerals 24, 141, 145, 167, 170
manganite 32, 85
Manorgill mine *45*
marble 14
marcasite 30, 55, 56, 78, 99
Margaret mine *83*, 84, 90
'marine bands' 13
Marl Slate 15
Maryport-Gilcrux fault 24
Meal Bank Quarry *72-73*
melanterite 55
Mell Fell Conglomerate 10
Melmerby Fell 48; Scar Limestone 14, 49, 66

Collectors and Collections

Mineral Dealers

Furnace, Anthony 112
Gilmore, Elizabeth 109
Gilmores of Alston 108,
 109, 174
Gilmore, Patrick 109
Gilmore, Peter 94, 109
Graham, Joseph 101,
 112
Graves, F. W. 111
Graves, John 102, 103,
 108, *110*, 110-11,
 113, 119, 127, 129,
 130, 132, 136, 140,
 141, 145, 146, 149,
 161, 164, 169, 172
Greenbank, Lindsay and
 Pat 105
Greenip junior, William
 110
Greenip senior, William
 110, 112
Henson, S. 152, 153,
 167
Humphrey, George 100
Hutchinson, Arthur 111
J. R. Gregory & Co.
 148
Kahn, Lazard 110
Pattinson, John 112
Stevens, J. C. 147
Talling, R. 103, 104,
 145
Tennant, Mr 103
UK Mining Ventures
 105
Williams, Scott J. 112
Wright junior, Bryce 112
Wright senior, Bryce
 McMurdo 103,
 108, 111-12, 119,
 168
Wright, Charles H. 111

Wright, Elizabeth 112
Wright, Maria ('Miss
 Wright') 111

Personnel (General)

Angus, John 102
Attwood, Charles 48
Auden, W. H. 106
Bathurst, Charles (C. B.
 Mines) 72-73
Bathurst, Dr J. 72
Beaumont Esq., Walter
 94-95; Beaumont
 family 94-95
Bell, W. W. 172
Birley, Miss Caroline
 150
Blackett, Sir William 44
Blight, Percy 101, 102,
 174
Bolton, Lord 72
Boulton, Matthew 123
Bournon, Count 117
Briethaupt, A. 124
Burlington, Earl of 80
Carlisle, Earl of 55, 62
Cleveland, Duchess of
 97
Constable, John 105
Derwentwater, Earl of
 44, 72
Devonshire, Dukes of
 72, 97, 80
Donaldson family 101
Dutton, Sir Ralph 98
Elizabeth I, Queen 24,
 38
Embrey, P. G. 138
Fletcher, Dr Lazurus
 103, 111

Gregory, J. 125
Hamilton, Charles 97
Heller, Joseph 102
Hochstetter, Daniel 36,
 38, 39
Hutton, James 98
Jamieson, James 124
Laverick, Lancelot 101
Lowry, L. S. 106
Lyell, Sir Charles 98
Martindale, Mr 101
Millican, H. 101, 155
Murray, Dr 174
Pattinson, Hugh Lee 46
 (*and see* Pattinson
 pans)
Potter, Beatrix 2
Rawnsley, Canon 2
Reed, Thomas 152,
 158, 161
Richardson, Joseph 10
Robinson, J. 102
Rutherford, Mr 95
Sedgwick, Adam 106,
 108, 111
Smyth, W. W. 161
Sopwith, Thomas 58,
 94, 95
Spencer, L. J. 111
Thomson, Thomas 124
Thompson, J. 136
Turner, Joseph Mallord
 105
Wainwright, Alfred 113
Ward, Clifton 110
Watt junior, J. 123
Wedgwood, Josiah 123
Wharton, Lord 72
Withering, Dr William
 123
Wordsworth, William
 2, 105-106

Associated Locations and Organisations

Beamish Museum,
 County Durham 114
British Geological Survey
 64, 82
British Museum (Natural
 History) 100, 103
Cambridge, University
 of 98
Cleveland Museum of
 Natural History 113
Hancock Museum 113
Kendal Museum 10
Kendal Natural History
 and Scientific Society
 108
Keswick Literary and
 Natural History
 Society 109, 112
Keswick Mining
 Museum 10, 11, 20,
 36, 114
Killhope Lead Mining
 Museum 44-45, 59,
 64, 66, 67, 95, 96,
 97, 105, 106, *114*;
 Wheel 106
Manchester University
 Museum 112
Museum House
 (Keswick) (1784) 107
Museum of Practical
 Geology 102
National Museums
 Scotland 112
National Museum
 Wales, 104, 112
National Parks (England)
 1
National Trust 2

Natural History Museum
 (and collections) 24,
 49, 77, 79, 80, 94,
 100, 101, 102, 103,
 108, 109, 111, 112,
 120, 124, 136, 138
 and *passim*
Nenthead Mines Heri-
 tage Centre 55, 114
North Pennines Heritage
 Trust 53, 114
Nottingham Castle
 Museum and Art
 Gallery 98
private museums *106*,
 106-108, *110*
Royal Cornwall
 Museum 99
Ruskin Gallery, Sheffield
 104
Ruskin Museum,
 Coniston 10, 36, 43,
 103, 104
Sedgwick Museum of
 Earth Sciences
 Cambridge 99
Sunderland Museum
 106, 113
Threlkeld Quarry and
 Mining Museum 114
Tullie House Museum
 (Carlisle) 104, 113
Upper Wharfedale
 Museum 80
Weardale Museum (St
 John's Chapel) 94,
 114
World Museum
 Liverpool 112
Yorkshire Museum 113